Engineering Contracts

Engineering Contracts

Robert Ribeiro MA LL.M. Ph.D. Barrister

Butterworth-Heinemann
Linacre House, Jordan Hill, Oxford OX2 8DP
A division of Reed Educational and Professional Publishing Ltd

 A member of the Reed Elsevier plc group

OXFORD BOSTON JOHANNESBURG
MELBOURNE NEW DELHI SINGAPORE

First published 1996

© Robert Ribeiro 1996

All rights reserved. No part of this publication
may be reproduced in any material form (including
photocopying or storing in any medium by electronic
means and whether or not transiently or incidentally
to some other use of this publication) without the
written permission of the copyright holder except
in accordance with the provisions of the Copyright,
Designs and Patents Act 1988 or under the terms of a
licence issued by the Copyright Licensing Agency Ltd,
90 Tottenham Court Road, London, England W1P 9HE.
Applications for the copyright holder's written permission
to reproduce any part of this publication should be addressed
to the publishers

British Library Cataloguing in Publication Data
Ribeiro, Robert
   Engineering contracts
   1. Engineering contracts – England   2. Engineering contracts –
   Law and legislation – England
   I.Title
   344.2'0378'62

ISBN 0 7506 2498 1

Composition by Genesis Typesetting, Rochester, Kent
Printed and bound in Great Britain by Hartnolls Ltd. Bodmin, Cornwall

Contents

Preface	vii
Acknowledgements	ix
1 Planning and making contracts	1
Negotiating skills	1
Some legal questions answered	12
2 Structuring contracts	19
The parties and their representatives	19
Obligations of the parties	23
Tests, taking-over, acceptance and rejection	31
Some legal questions answered	36
3 Price and payment	39
Price	39
Payment	42
Some legal questions answered	47
4 Terms about risk and delivery	55
Risk	55
Terms about delivery	57
The meaning of times for delivery or performance	62
Some legal questions answered	67
5 Progress in engineering contracts	74
The programme	74
Force majeure	75
Claims for additional payment	77

	Suspension and termination of engineering contracts	80
	Some legal questions answered	83
6	**Quality and fitness for purpose**	**89**
	The legal and commercial kaleidoscope	89
	Defects liability and express warranties	96
7	**Liabilities, exclusions and indemnities**	**106**
	Different kinds of liability	106
	Negligence	108
	Product liability	112
	Limits of liability	116
	Some legal questions answered	121
8	**Ownership of goods and intellectual property rights**	**125**
	Ownership of goods and materials	125
	Intellectual property rights	131
	Some legal questions answered	135
9	**Multipartite arrangements**	**138**
	Agency, sub-contracting, and free-issue	138
	The chain of responsibility	142
	Some legal questions answered	146
10	**Negotiating legal and financial matters**	**152**
	Performance bonds and guarantees	152
	Insurance and engineering contracts	160
	Arbitration clauses, and the duration of liability	163
	The future	171

Appendix 1	175
Designing and structuring an engineering contract	175
Appendix 2	178
Statutes and other legislation	178
Appendix 3	179
Cases	179
Appendix 4	183
List of engineering institutions and their engineering contracts	183
Glossary	185
Select Bibliography	190
Index	191

Preface

Engineering Contracts is intended for those who wish to acquire skills in drafting, negotiating or working with commercial and engineering contracts. It aims to bring a different approach to the subject: the traditional work on the law of contract, with its emphasis on law as a series of themes and rules, can often be frustrating for the reader who is in search of legal solutions to commercial problems. To the commercial manager or the engineer, as well as to members of many other professions, the crucial questions are about how to plan, negotiate, draft, document, interpret, perform and obtain commercial benefits from contracts. These are the matters with which this book deals, and it is hoped, in particular, that the reader will find helpful the 'legal questions answered' sections that I have incorporated into most of the chapters.

I have given the expression 'engineering contracts' the widest possible meaning, and for the purposes of illustration of points, I have drawn cases from the fields of mechanical, electrical, chemical, electronic and civil engineering, as well as from building and construction contracts. Many of the illustrative cases are about sales of goods (and some lie outside the fields of engineering or manufacture altogether), because they provide the only appropriate examples of an important legal point. Commonwealth cases, as well as English and Scottish cases have been noted and the impact of European Community law has occasionally been alluded to. However, the book remains primarily a book about English commercial law. The law stated is that in force on 1 March 1996.

Acknowledgements

I would like to acknowledge the help, facilities and access to materials provided by the many companies with which I have come into contact during the preparation and writing of this book, as well as the many occasions on which their staff have discussed with me the issues which have formed the subject matter of this book, and have been ready to offer valuable suggestions.

My thanks are due to Michael Forster of Butterworth-Heinemann, for the encouragement he has given me in developing this book.

Most of all I would like to thank Aileen Ribeiro for her advice, help, and patience during the year of writing.

Robert Ribeiro
London
1996

1 Planning and making contracts

You can never plan the future by the past

Edmund Burke

Negotiating skills

Nothing stands still in the world of commerce and new laws and interpretations of existing laws relating to it arise continuously. New issues surface; there are shifts in the emphasis or focal point of the law relating to commercial and engineering contracts. Side by side with familiar points of English law there are new and sometimes difficult questions raised by European Community law. Side by side with the customary 'arms length' approach to commercial contracts, there are new relationships developing: the joint venture and the shared enterprise; the cooperative or non-adversarial approach. These, where they exist, call for new and innovative skills and ideas and uses of legal methods. Everything that is stated in this chapter must be taken as serving one primary purpose: that of using the laws that we have and the choices that they present us with to serve commercial ends.

Engineering contracts share many features in common with other types of contract. They are often distinguished by their size, complexity, technical content and time-scales, but in their essential objectives they tend to reflect a purchaser's point of view, which is the need to obtain from a structure, equipment, or hardware or software, a required standard of performance at the required time and at the required price, as well as to reflect a contractor's view of a profitable transaction leading to a satisfactory conclusion, and trouble-free relationship between the parties.

To negotiate such contracts calls for a wide range of skills: legal, technical, and financial, as well as skills in basic drafting and comprehension. Management skills are needed to evaluate the risks and rewards, and to negotiate to achieve the required balance. This may, on occasions, be carried out by one highly skilled individual, or, more commonly, there will be team involvement and support, often with the aid of forms of contract and other documents supplied by professional institutions.

Engineering contracts have little in common with gambling, and if one strategic end can be described as paramount, it must be that, as far as possible, the outcome of such a contract must be predictable. With this in mind, we can identify four different philosophies or functions relating to engineering contracts:

1 planning;
2 achievement of mutual understanding and expectations;
3 financial management and controls;
4 risk management and allocation.

These disciplines could be applied to a contract by a person with no specialist legal knowledge or training, but there would be obvious deficiencies in taking such an approach. Many of the choices to be made are commercial and technical, but legal understanding, as well as commercial awareness must guide these choices. The full legal implications of each step should be well understood by every person involved in the negotiations. For those who approach an engineering contract as part of a commercial plan, it is essential to bear in mind that the plan must exist at all times within a legal framework, and that failure to appreciate this may result in a serious flaw in an essential part of the plan. In the well-known engineering case of *British Steel Corp.* v. *Cleveland Bridge and Engineering Co. Ltd* (1981, reported in 1984), a failure by a purchaser to appreciate that a 'letter of intent' was not contractual and could not give rise to a binding programme of delivery, or to legal liability for delay, had unfortunate consequences which might well have been avoided if the four philosophies which are mentioned in this chapter and reiterated throughout this work had been applied.

A contract is a part of a commercial plan, but it may also be defined as a legally binding agreement. As a matter of planning, there may be preliminary inquiries of a commercial and technical nature, followed by a specification of a product or service. A programme may indicate dates or periods of time for manufacture, delivery, construction (where relevant) and the carrying out of a series of tests. Terms of payment and other terms of contract will be mentioned, and contractors will be invited to quote or to tender. At this stage there is, of course, no contract and the parties may well have differing views as to what the terms of contract should be. The outcome will depend upon how the negotiations are carried out, and upon the conclusion to those negotiations.

Parties to commercial negotiations should at all times bear in mind that at the end of these negotiations any of four outcomes is possible. There may be:

1 a contract based upon terms and conditions proposed by the purchaser;
2 a contract based upon the terms and conditions proposed by the contractor;
3 a contract based upon a 'neutral' set of terms and conditions, or an agreed compromise;
4 no contract.

Where negotiations result in there being no contract, this may lead to further negotiations, or it may mean that matters do not proceed any further. The main risk at this stage, however, is that one of the parties may not appreciate the position and may assume that a contract exists. For this reason it is important that those concerned should know what the essential requirements for a binding agreement are, as well as the different ways in which contracts may be formed.

The essential requirements for binding agreements

A contract requires an *offer* (for example, a tender, quotation or purchase order) to be made by one party and to be accepted by the other party. The *acceptance* must take place before the offer lapses or before it is withdrawn. If a period of validity is stated in the offer it will lapse at the end of this period. Otherwise it lapses after a reasonable time, if not accepted. The parties must have *capacity* to contract, and the objects of the contract must be *legal*. The parties to a contract must be shown to have *intended to create legal relations*; this intention will usually be inferred from the circumstances and from the wording used. Suitable wording may negative any intention at a particular stage of negotiations to create legal relations, a point which may be put to good use when parties make certain communications, such as letters of intent, without necessarily wishing to enter into a binding contract at that particular time. Letters of intent can certainly have legal consequences, as we shall see in due course, but those consequences are not necessarily contractual.

A further, important, requirement for a binding contract is that each party should promise or undertake something of value in return for the promise of the other. This is known as the requirement of *consideration*, and consideration must exist in one form or another in every contract with the exception of those contracts which are made by deed. If a contract is made by deed, which is to say in writing, and signed and sealed or otherwise expressed to be a deed, then there is no requirement of any consideration in return for the undertaking given in that deed. In practice, in building and engineering contracts, wherever deeds are used there is usually consideration as well, and the purpose of the deed is not to avoid the rule of consideration, but to provide other advantages which will be discussed later in this work. Consideration need not be equal on both sides, and a contract to sell a valuable piece of equipment for a nominal sum such as £5.00 will satisfy the requirement of consideration. In summary the legal requirements of a commercial contract are normally:

- Offer
- Acceptance
- Capacity
- Legality
- Intention to create legal relations
- Consideration

It will be noted that formality, in the sense of a written document and a signature, is not normally one of the requirements of a binding contract. This has advantages, in so far as an ordinary commercial contract may be made by the most convenient method, including oral and electronic communications. The disadvantages of making engineering contracts by less formal methods are not legal but administrative, and usually relate to problems of identification of the terms of the agreement, or of proof that a contract actually existed in the first place. It is for these reasons, and not because of any legal requirement, that engineering contracts are usually brought into being by means of carefully prepared and signed documents.

Some types of contract are required under English law to be made in writing and signed. Of these, the most common are contracts for the sale or other disposition of an interest in land. These are now governed by the Law of Property (Miscellaneous Provisions) Act 1989, Section 2 of which sets out the details of the requirement that such contracts must be made in writing. As well as these, regulated Consumer Credit Agreements are, with certain exceptions, required to be made in writing and in prescribed form. Contracts of guarantee, in the sense of one party promising to answer for the debt or default of another person or company, must be evidenced in writing in order to be enforceable. As far as engineering contracts are concerned, this has possible implications in so far as performance or payment may be the subject of a guarantee given by a third party, such as a bank or a parent company. In practice, the party receiving and relying upon the guarantee would always insist that it be in writing and signed by the guarantor – even if this were not required by law.

Simple contracts: evaluating different types of communication

The communications which take place in business transactions, whether they are day to day transactions or major commercial projects, often present practical difficulties. It is not always clear what a document is

Planning and making contracts 5

meant to signify in legal terms when it is written in a format or language reflecting the business practice of a particular company or organization, or sector of industry. When we try to relate it to the legal requirements of contracts, it will be seen that a great deal depends upon the clarity and sequence of communications between the parties. Negotiating skills will be considerably enhanced if we bear this in mind and try to relate business practice to the concepts of offer and acceptance.

In a typical pattern of negotiations about an engineering project there may be communications of the following kind (other patterns of communication are, of course, possible).

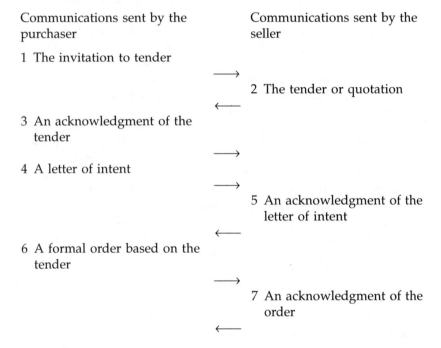

The legal issue is almost invariably one of when an offer has been made, and when it has been accepted so as to form a binding contract. There is also the question of identifying the terms of the contract.

It is useful, in forming an analysis of such negotiations, to keep in mind some basic categories into which communications may fall.

1 *Making an inquiry, and giving information*: these do not amount to offers or acceptances. People are entitled to ask for factual information, or to give such information, without necessarily intending it to have legal consequences. Information given can at a later stage (if and when a contract is made) have legal consequences if it was relied upon and

turns out to be incorrect – under the law of misrepresentation. But unless a contract comes into being there can be no legal relationship arising out of purely factual information, such as catalogues or price lists or technical information about products.
2. *Invitations to treat*: these may be made by either party to a prospective contract. One person is, in effect, *inviting* an offer from the other. An invitation to quote or to tender is an *invitation* to treat. It is not an offer, so for a contract to arise it must be followed by an offer and an acceptance. In general, an invitation to treat does not carry legal consequences for either party, but there may be exceptions to this. In *Blackpool & Fylde Aero Club* v. *Blackpool Borough Council* (1990), the Court of Appeal held that an invitation to tender can give rise to a binding obligation: if the language and circumstances of the invitation to tender make it clear that one party is undertaking an obligation to consider all conforming tenders, then damages may be awarded for a failure to consider one such tender along with the others.
3. *Offers*: the best definition of an offer is a communication by one person to another, which is intended to constitute an offer, and which is capable of being accepted. In engineering contracts the most common types of offer are tenders and quotations, which are made by prospective sellers and contractors. Intending customers may also make offers, in the form of offers to purchase. Purchase orders are therefore capable of being either offers or acceptances of offers, depending upon what has gone before. This is one of the reasons why the most careful analysis of the whole sequence of communications is needed. Another possible complication is the fact that not every quotation is necessarily an offer. It is often thought that a quotation is an offer and an estimate is not, but in reality there is no such rule. Whether a document is called a quotation or an estimate is not the true test of its legal meaning: the legal intention is to be found in its detailed wording and purport. A disclaimer of any intention to make an offer will prevent a quotation from being an offer. In the absence of a disclaimer or of any qualifying words, an estimate is capable of being an offer.
4. *Acceptance*: to form a contract an offer must have been accepted. The acceptance must be unconditional, which is to say that it must only accept exactly what is offered and must not purport to accept something different, or in any way alter the terms of the offer. If it does alter the terms of the offer, it will not amount to an acceptance at all, but will be construed as a *counter-offer*. A counter-offer is, in law, equivalent to a rejection of the original offer, replacing it with a new offer by the party making the counter-offer. In a major engineering case, *Trollope and Colls Ltd and Holland and Hannen and Cubitts Ltd, trading as Nuclear Civil Constructors (a firm)* v. *Atomic Power Constructions Ltd* (1962), it was held

by the judge that 'the counter-offer kills the original offer'. So strong is this rule, that a person who has made a counter-offer cannot change his mind and accept the original offer, unless the other party is prepared to renew the original offer.

The timing of the acceptance

This can sometimes be of practical importance, since the rule is that an acceptance, if it is to be valid, must be effectively made while the offer still stands. In engineering contracts, most tenders and quotations are expressly stated by the party making them to be open for a limited period, such as sixty or ninety days. At the end of this period, the offer automatically lapses. But the person making an offer may, in fact, revoke it at any time before it has been accepted. The fact that an offer has been stated to be open for acceptance for a given period does not prevent that offer from being withdrawn earlier, although as a matter of commercial credibility contractors will not withdraw offers unless there are particularly compelling reasons for doing so. To revoke or withdraw an offer, the notice of revocation must be communicated to the person who has received the offer. If communications cross, it becomes crucial to know at what point they become effective. The rules which follow are the rules in English law. They are not necessarily the same rules that apply in other countries, so special care is needed for overseas contracts and international communications.

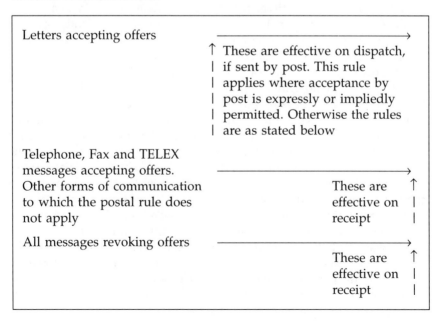

Offer and counter-offer

A contract is said, as a principle of law, to come into being as soon as an offer made by one party is accepted by the other. The terms are then the terms of the offer. But if the purported acceptance is not unconditional, then it is not an acceptance at all; it is in fact a counter-offer. An important part of the art of negotiating contracts is to be able to identify a counter-offer, and to decide how to respond to it. In many areas of business, particularly as far as sales and purchases of goods and materials are concerned, counter-offers follow offers as a matter of course. This is because so many businesses have forms of purchase order and of acknowledgement of order which refer to standard conditions printed on the back of those forms. Not unnaturally, sales staff and purchasing staff make use of such forms in responding to each other, and when they do so the chances are that the conditions mentioned on the forms will differ from, or contradict, conditions already proposed by the other party. This may happen early on in the negotiations, or it may even occur at the very last moment, thus wiping out many months of careful preparation by the parties. All is not lost, however, provided that those concerned are aware that a counter-offer has been made, and have plans or a system in place to decide what to do next.

It helps to remember the basic rule, which is that there is *no contract until the parties have reached agreement*. This, in major projects where a proper Memorandum of Agreement is to be signed, will normally mean that a contract will not exist until the terms and conditions of contract have been finalized and agreed upon. It is unlikely that the parties will fail to do this, although such a failure did occur in the case of *British Steel Corp. v. Cleveland Bridge and Engineering Co. Ltd* (1981). In this case British Steel Corp. (BSC) successfuly tendered for the production of cast steel nodes which were to be used by Cleveland Bridge and Engineering Co. Ltd (CBE) in a project in Saudi Arabia. The project required 137 cast-steel nodes for the centre of a steel frame for a bank in Saudi Arabia. The nodes were of unique specification, and CBE – perhaps foreseeing difficulties in delivery – sent BSC a letter of intent requesting BSC to 'proceed immediately with the works pending the preparation and issuing to you of the official form of sub-contract'. BSC processed this instruction and began preparations on the work. The conditions of contract had not yet been agreed, and in particular a formal quotation had not yet been issued, the tender price being merely an estimate based on incomplete information. The conditions put forward by CBE provided for unlimited liability for loss arising from late delivery. These conditions were unacceptable to BSC. However, throughout the period of the negotiations, production went ahead and, in the event, BSC delivered all but one of the 137 nodes, the last being held back to ensure that payment would be made.

At this stage of this involved story we may pause and reflect that there are a number of negotiating lessons to be learned here. Firstly the letter of intent was not a contract because the situation of offer and counter-offer had not yet been resolved. Neither party had accepted the terms proposed by the other, and a compromise had not as yet been reached. Mr Justice Robert Goff stated in the Commercial Court that: 'There can be no hard and fast answer to the question whether a letter of intent will give rise to a binding agreement: everything must depend upon the circumstances of the particular case.' He also added: 'The real difficulty is to be found in the factual matrix of the transaction, and in particular the fact that work was being done *pending* a formal sub-contract the terms of which were still in a state of negotiation.' Clearly, there is a possibility of legal confusion if such a situation is allowed to arise, and such a situation should either not be permitted at all, or should be carefully controlled. When BSC held back the final node, they were doubtless aware of these issues, but what they did was also capable of creating complications and difficulties.

Secondly, a review of the facts and of the time-span of this case raises the question as to why a contract was not drawn up and concluded as quickly as possible. One may only speculate on the reasons for this, and suggest that the important thing is for the parties to meet or communicate frequently, and never lose sight of the need to bring the negotiations to a satisfactory conclusion, and to create and record legally binding contracts. It is the contract alone that brings the required element of certainty and definition into the business venture.

In the event, the last of the steel nodes became trapped in a strike in the steel industry with the result that delivery of it was far later than the intended schedule. BSC submitted its invoice for the price of the goods delivered but was confronted with a counterclaim by CBE for damages for late delivery. The missing node had caused a delay in construction which resulted in losses to CBE of £896 715.38; as a consequence of this, by way of set-off and counterclaim, CBE brought an action for the balance of £666 882.68, the invoiced price being £229 832.70.

The court held that since there was no contract in this instance, there could be no liability to deliver goods within any particular time and, therefore, BSC could not be liable for any delays caused to CBE. But although there was no contract, there *was* a liability on the part of CBE to pay for the nodes. The reason for this is that liability to pay need not arise out of a contract, although in most cases a contract will exist. If, however, goods are delivered or services rendered, which the purchaser expressly asked for, and if there is no contract, the purchaser's liability to pay will arise out of the law of 'quasi-contract' or 'restitution'. This law is based simply upon principles of equity or justice, and the liability of the purchaser is to pay a reasonable sum to compensate for the value of the goods or services. In the actual case of BSC and CBE, the sum in question was agreed by the parties during the course of the hearing, and was in fact the amount actually claimed by BSC.

A further point about counter-offers needs to be made: offers and counter-offers can be accepted by *implication* or by *conduct*. In the BSC/CBE case this possibility was discussed in court, but was ruled out because the course of the negotiations made it very clear that the parties disagreed strongly about the terms that were under discussion, and that further discussions were contemplated. However, in some instances, although there may be a conflict between the terms and conditions put forward by the parties, it may be possible to infer from the way in which the parties have behaved or communicated with each other that one of the parties has backed down and has been content to allow the contract to be made according to the terms and conditions put forward by the other.

Where the courts are prepared to find an implied acceptance of an offer, or an acceptance by conduct, it will usually be on the basis that there was no response to – or objection to – the *last* set of terms and conditions proposed, and the party receiving this final offer did something that amounted to consent to what was offered. At this point the contract will have come into being, and in many instances no formal communication, and sometimes not even a written document, will be necessary. What is, however, needed is an action which amounts to a clear communication of assent, so complete silence and inactivity on the part of the person receiving the offer can never amount to an acceptance.

In the case of *Re Bond Worth Ltd* (1979), Bond Worth Ltd was the purchaser, and had ordered goods from Monsanto Ltd, on printed order forms with conditions of purchase on the back. Monsanto Ltd sent a series of twenty-nine confirmation notes, since they intended to make delivery at a number of different times. Each of the confirmation notes had on the back the conditions of sale of Monsanto Ltd, which had also been drawn to the attention of the Company Secretary of Bond Worth Ltd. Bond Worth Ltd did not make any express objection to the conditions of sale of Monsanto Ltd, and goods were delivered on a number of occasions, until a legal dispute arose concerning payment and title to the goods. It was important to know which conditions were applicable, since this was capable of affecting the position as to title. Mr Justice Slade, in the Chancery Division of the High Court, held that the final offer (which was in fact a counter-offer) had been made by Monsanto Ltd on each occasion that a confirmation note was followed by a delivery of the goods. There were in fact twenty-nine different contracts, and each was made at the point of delivery and acceptance of the goods. Although Bond Worth Ltd did not respond to the counter-offers of Monsanto Ltd, the taking of delivery, after notice of and knowledge of the terms of each counter-offer, amounted to acceptance by conduct.

Thus, it can be said that where there is no evidence of actual objection to the terms and conditions offered, a party may through its conduct be bound by the final set of terms and conditions that have been fairly and reasonably brought to its attention. This has, not unnaturally, caused a certain anxiety in business circles, since it is possible that a person of no great legal or contractual expertise might be put into a situation where a contract is made by conduct alone. The most obvious example is where a person signs for a consignment of goods. All that can be said is that managers should be aware that this can happen, and appropriate steps should be taken to control the process of offer and acceptance, by responding to all counter-offers, and by proper designation of authority to contract. Apart from this, the law would be applied on a case-by-case basis, exactly according to the particular circumstances. In a number of cases the courts have been prepared to look at whether or not the last set of terms and conditions was fairly and reasonably drawn to the attention of the other party.

In the Scottish case of *Grayston Plant Ltd* v. *Plean Precast Ltd* (1976), where a contract of hire was made, and the conditions of contract were sent to the customer after a telephone conversation ordering the hired goods, it was held that the written conditions were not applicable to the contract. This is because they were not known to the customer and had not actually been mentioned in the telephone conversation and, therefore, it could not be said that they had been properly brought to the attention of the customer. When the conditions did arrive, by post, the contract had already been made, so the customer could not be said to have assented to the conditions.

In the case of *Interfoto Picture Library Ltd* v. *Stiletto Visual Programmes Ltd* (1988), the requirement of *fairness* in bringing terms and conditions to the attention of the other party was specifically underlined by the Court of Appeal. One of the questions most frequently asked is whether or not a seller can slip a new set of conditions into a contract at the last moment, by means of delivery note or advice note. In most cases, such documents (including invoices) will have no contractual effect, since they are usually issued after the contract has been made: but what if there is still no contract at the time of delivery? The advice note could also be the final offer. The problem for the purchaser is that, at this moment, the conditions on such a document are unlikely to be read by the people taking delivery of the goods. In the case in question, a bag of transparencies was delivered with a delivery note which contained the words 'CONDITIONS' in capitals. It then set out the conditions, which were related to the contract of hire, and imposed a charge of £5.00 per day for each transparency retained beyond an initial fourteen-day period. The customer was not aware of this condition, and the relevance of the fourteen-day period only became apparent when an invoice for a far larger sum than had been contemplated was issued by the supplier. The Court held that the condition in question had not been accepted by the customer, since it was an unusual condition which would not normally be known to a customer, and had not been fairly and reasonably drawn to the attention of that customer. Lord Justice Bingham stressed the requirement of fair and open dealing, and held that the customer was only liable to pay the normal hire charges, but not the abnormal charges relating to the exceeding of the fourteen-day period.

Some legal questions answered

In this second part of this chapter we will look at some of the questions that those involved in negotiating engineering contracts are most likely to ask concerning legal issues; we will consider how the law has approached such questions.

Can a letter of intent ever have binding effect as a contract?

The answer is that it can, depending upon what is stated in the letter. Most of the letters of intent that are written contain qualifying words to the effect that the contract is still under discussion or subject to the issuing of an official order or letter of acceptance. Such qualifying words show clearly that there is no intention to create a contract at this stage. If a letter of intent had the features of an acceptance of a tender, and had no qualifying words, it would probably amount to a contractual document. It should be noted also that it is possible to write a letter of intent with a limited commitment which is contractually binding even if the remainder of the letter of intent is not. For example, a purchaser may state that the contract is subject to confirmation in an official order, but that the contractor is authorized by the letter of intent to undertake preparatory work up to a limited (and stated) value for which the purchaser undertakes to pay.

If work is commenced on the basis of a letter of intent, and if a contract is subsequently made between the parties concerning the same project, will the terms of the contract govern retrospectively?

The answer to this question is that it depends upon the intentions of the parties, which will have to be interpreted from the contract itself and from the letter of intent. It is possible for the parties to come to whatever arrangements they wish, either treating any preliminary work as part of the finalized contract or treating it separately. Financial consequences will possibly differ according to the arrangements chosen, so what is needed at all stages is freedom from ambiguity. In the case of *Trollope and Colls Ltd, etc., trading as Nuclear Civil Constructors (a firm)* v. *Atomic Power Constructions Ltd* (1962) millions of pounds depended upon the answer to the question of the retrospective effect of signing a contract. In February 1959 the parties began a long series of discussions about the terms and conditions and a specification; in June 1959 work began on the basis of a letter of intent; and in April 1960 the contract was made.

Trollope and Colls Ltd and Holland and Hannen and Cubitts Ltd would probably have received a higher sum in payment for the work done between June 1959 and April 1960 if it had not been governed by the terms of the contract eventually made. Had they realized this in time, they might have taken a different approach to the finalization of the contract. It would have been perfectly possible to have had two separate contracts with different terms of payment governing different phases. However, the judge held that it was correct to infer that the parties intended the terms of the contract to apply retrospectively.

How do the parties to a complex engineering contract identify all the documents relevant to the contract?

They normally conclude the making of the contract with a letter of acceptance, or Memorandum of Agreement. Either of these may list all the documents which the parties wish to be in the contract.

Should there be an order of precedence of such documents?

Ideally there should, in case a conflict between documents should arise. (Further details of this are set out in the Appendix.)

What is the point of an 'entire agreement' clause?

This is a clause which makes it clear that the documents listed as being the contract documents constitute the entire agreement between the parties, and that the parties are not relying upon any other document, or upon any oral terms, and that the documents listed are intended to supersede all other documents and negotiations. The point of the clause is to bring as much certainty as is possible to the formation of the contract. In the case of *McGrath* v. *Shah* (1989), a case about purchase of land, the High Court held that an 'entire agreement' clause is a perfectly fair and legitimate negotiating procedure. It is not unfair, because, if properly written, it is simply advising the parties as to where to look for all the terms of the agreement.

Can parties to a contract ever look outside the terms of a written agreement?

The issue is always one of the intention of the parties. If there are oral discussions, followed subsequently by a written agreement, it is possible to infer that the written documents reflect the whole of the agreement. This will be particularly so if there is an 'entire agreement' clause. But it is also possible for a contract to be made partly orally and partly in writing. Or it could even be that the oral agreement constitutes one contract and the written agreement constitutes another. This was the position in the case of *J. Evans and Son (Portsmouth) Ltd* v. *Andrea Merzario Ltd* (1976). This case shows once again that care is needed in any negotiating procedures, and that meetings should either be designated as informal, or else should be minuted and designated as contract documents. J. Evans and Son Ltd, importers of machines from Italy, had discussions with the general manager of the London offices of Andrea Merzario Ltd, regarding the carriage of an injection moulding machine in a container. The risk of possible damage by sea spray had been identified from previous occasions, and J. Evans and Son Ltd, as purchasers, insisted that the machine should be shipped in a container under deck. The assurance received was purely oral, and was not recorded in any of the contract documents. On the basis of this assurance, J. Evans and Son Ltd signed a contract of carriage. Some of the printed terms of this contract of carriage were, in fact, inconsistent with the oral assurance. By an oversight the container was shipped on deck, and it fell overboard and was lost. J. Evans and Son Ltd claimed for damages against Andrea Merzario Ltd, alleging that there had been a breach of contract which caused the loss of the machine. As the written terms of the contract of carriage limited the liability of Andrea Merzario Ltd, as well as giving them complete freedom as to where the goods were to be placed, it was essential for J. Evans and Son Ltd to be allowed to look outside the written documents, and to give evidence as to the oral assurance. The Court of Appeal allowed J. Evans and Son Ltd to bring evidence to show that such an assurance had been relied upon, and it held that the assurance was contractually binding. Accordingly, J. Evans and Son Ltd was entitled to damages for breach of contract.

What is a 'course of dealing' between two parties to a contract?

A 'course of dealing' only arises when the parties to a contract do not set out the terms and conditions prior to the agreement. With major contracts this situation is unlikely to arise, but with smaller engineering contracts or contracts of carriage, or other small commercial contracts it is quite possible that two parties who deal with each other frequently will, after a while, make the assumption that the details of the terms and the conditions (other than price and specification) need not be discussed, since they will already be known to both parties. Parties may, of course, make contracts by shorthand forms of reference, such as to the JCT forms, or to BEAMA or FIDIC or similar standard conditions. (These will be detailed later in this work.) Parties may also refer simply to the standard terms of one of the parties. In *Smith v. South Wales Switchgear Ltd* (1978) a company made a purchase order stating in writing that the order was 'subject to our General Conditions of Contract, obtainable on request', and this was held to be sufficient to incorporate into the contract the latest revision of the conditions at the time of the placing of the order.

Sometimes, however, the parties do not even wish to go to this much trouble, and the barest amount of information may be transmitted between the two parties. It is not, by any means, the ideal way to make a contract, but sometimes urgency and commercial pressures are overwhelming and short cuts are taken. Provided that at some point in the past, with reasonable consistency, and on a reasonable number of occasions, the same two parties have used the same terms and conditions in practice, it may be possible to argue that an established 'course of dealing' has come into being. In the case of *Circle Freight International Ltd v. Medeast Gulf Exports* (1988) the Court of Appeal stated that it is not necessary that contract conditions should always be set out, provided that adequate notice is given identifying the conditions, and making it clear that they are available on request, as long as the terms are not particularly onerous or unusual. In this particular case, the parties were commercial companies. There had been a course of dealing in which at least eleven invoices had been issued, and although an invoice is not part of a contract, the terms printed on each invoice served to give the customer notice of the standard conditions of the forwarding agents. The conditions in question were the standard conditions of the Institute of Freight Forwarders (IFF) and were in common use and were not particularly onerous. The customer, who had never read the conditions printed on the invoices contended that these conditions were not incorporated into subsequent contracts, but the Court of Appeal held that the conduct of the customer, in continuing the course of business after at least eleven notices of the terms would have led – and did lead – the freight forwarders to believe that their terms had been accepted by the customers.

Can a contract be made by facsimile and other electronic methods?

The answer is that there is no reason why not as a face-to-face oral offer and acceptance is enough to constitute a valid contract, and offer and acceptance by ordinary telephone conversation is also sufficient to constitute a contract. The difficulty in such cases is often one of proof of exactly what the terms and conditions were, so that it is unlikely that oral methods of contracting will be used for complex engineering contracts. Facsimile simply goes one step further, since the telephone is used but the message arrives in the form of a copy of a document. It would be unwise to rely upon a fax where an original document is required: a copy is only secondary evidence. But it is far superior to an oral contract, so there can be no real grounds for objection to a fax message, subject to the following provisions: it must be fully legible, because if it is not, then there could be doubt as to whether the parties were in agreement; it should be properly numbered and dated; and it should be clear who is the originator of the fax message, since the question of authority may be raised. It should not refer to words on the reverse of the original document, but if such references are to be made, the references should be to page numbers, and all the pages should be sent. One problem which is bound to be raised sooner or later is that of a fax message which is sent out of office hours: will it be an effective acceptance or counter-offer, and if so, at what moment? All that can be said is that the courts will have to deal with such complications on a case-by-case basis.

Telex communications were considered by the House of Lords in 1982. The issue before them was at precisely what point in a series of negotiations a contract had been formed, if at all The buyers, in the case of *Brinkibon Ltd* v. *Stahag Stahl und Stahlwarenhandelsgesellschaft MbH* (1983), contended that a contract had been made, and that the sellers were liable for damages for failure to deliver. The sellers argued that the communications did not amount to a contract, or that if they did amount to a contract, then the contract was formed in such a way that it was subject to the jurisdiction of the Austrian Courts and not the English courts. The contract was for the supply of a quantity of mild steel bars. Negotiations had begun in April 1979. There were Telex messages from buyer to seller about the conditions of contract on 20 and 23 April 1979. The seller replied on 25 April 1979. The buyer replied on 26 April 1979, and by this time the parties were nearing agreement on terms about weight, price, C&F terms of shipment, payment by letter of credit, and a performance bond. The seller then sent a Telex on 3 May 1979, introducing new terms about freight charges and proposing that the performance bond be reduced from 5 per cent of the price to 3 per cent. The buyer agreed to this by Telex on 4 May 1979. Thereafter, the seller sent more Telex message suggesting that the arrangements were unworkable. The contract was not performed.

The House of Lords held that the communications did amount to a contract. The Telex message of 3 May 1979 could be regarded as the final counter-offer. The buyer accepted this unconditionally on 4 May 1979. This amounted to a contract, and any further Telex messages were of no effect for legal purposes. However, such methods of negotiation are not always satisfactory to the parties, since a series of Telex messages may omit important conditions of contract which may be needed in the event of a dispute. In this particular case, what was lacking was a provision that the contract was to be governed by English law and to be subject to the jurisdiction of the English courts, and so the House of Lords was unable to award damages to the buyer.

Are the minutes of a meeting contractually binding?

A great deal will depend upon what is intended by the parties, and upon the time at which the meeting takes place, and the purpose it is meant to serve. In the case of *Orion Insurance Co* v. *Sphere Drake Insurance* (1990) the High Court held that where the parties agree the terms of a contract at a meeting, of which a signed detailed minute is kept, there is a strong presumption that the minute contains enforceable contractual terms. In such a case the meeting would be before the contract is made, and the parties may approach it in a number of ways. They may, for example, decide that the minutes will be listed as a contract document; or they may refer to it in an exchange of letters afterwards; or they may, if they so wish, agree at the meeting itself that a contract has, at that moment, come into being. In any of these instances, it would be advisable that each party should carefully read the minutes and then sign. It is not good negotiating practice to rely upon unsigned minutes prepared by only one of the parties. To quote the words of Lord Justice Donaldson in the case of *Esmil Ltd* v. *Fairclough Engineering Ltd* (1981): 'It is a wise and elementary precaution to agree expressly upon the terms of a contract before undertaking its performance.'

What is the position if documents appear to be inconsistent or contradictory?

In major engineering contracts this problem should not arise, because the documents should be mutually agreed, and should state an order of precedence in the event of any conflict. But if the contract is made less formally, problems may arise. The *last* agreed set of terms and conditions should govern, but this simple rule does not deal with all complications which may arise. In the well-known case of *Butler Machine Tool Co. Ltd* v. *Ex-Cell-O Ltd* (1978), the seller sent a quotation to the buyer offering to sell a machine at a quoted price, but subject to conditions which stated that the price to be charged would be the price ruling at the date of delivery. The buyer placed an order, subject to the buyer's own conditions of purchase, which contained a tear-off slip to be returned by the seller. This tear-off slip stated: 'We accept your order on the Terms and Conditions stated thereon'.

The seller's sales department signed and returned this tear-off slip to the buyer, together with a letter stating that the buyer's order was being entered in accordance with the seller's original quotation. In the case which followed, the dispute concerned the price, and the question of whether it was adjustable – as the seller's terms stated – or fixed – as the buyer's terms stated. The Court of Appeal found in favour of the buyer, on the ground that the last communication, which contained both the tear-off slip and the letter, had to be read as one document, and the express acceptance of the buyer's conditions on the tear-off slip was decisive.

In the case of *Harvey* v. *Ventilorenfabrik Oelde* (1989), an interesting problem arose: the seller of two machines made out the agreement in duplicate on 'acknowledgment of order' forms. One form, kept by the buyer, had no conditions printed on the back. The other, signed by the buyer and returned to the seller, had on it conditions written in German. After the machines had been delivered, a dispute arose about quality, and the buyer wished to sue in England. The defence of the seller was that according to the signed conditions, the contract was subject to German law and German jurisdiction. However, the Court of Appeal in England found that the buyer was misled by the difference between the two sets of documents, and did not, in reality, assent to the incorporation of the jurisdiction clause. Both of these two cases last discussed show the extent to which care is needed in preparation and use of contract documents.

2 Structuring contracts

Be clear and the rest will follow

Napoleon Bonaparte

The parties and their representatives

An essential part of an engineering contract is the identification of all those who may be involved in it. This will be of more crucial importance in engineering contracts than it is in less complex commercial contracts, such as sales of goods. The reasons are, firstly, that virtually all engineering contracts envisage work being done for a purchaser whose rights, duties and powers will be exercised by an *engineer*, or *purchaser's representative*, or someone of similar designation. Secondly, in engineering contracts, the issue of sub-contracting, or sub-letting, and the framework within which this is to be permitted, is an important one. Thirdly, the long-term nature of some engineering contracts means that one cannot be certain that the parties will not change or wish to sell or part with their rights under the contract. A purchaser of a building or structure may, for example, wish to know whether or not he has the same rights in respect of defects as the original employer of the contractor who put up the building or structure. This will depend upon whether or not the original employer, when selling the building or structure, is entitled to *assign* his contractual rights and benefits under the contract.

The engineer

The expression 'the engineer' in the one most commonly used in engineering contracts to describe the person who is to act on behalf of the employer or purchaser for the purposes of the performance of the contract. Such persons may also be described in some engineering contracts as employer's representatives or as project managers. The choice of title is less important than that the person to fill the role should be named, and the meaning of the expression carefully defined. The

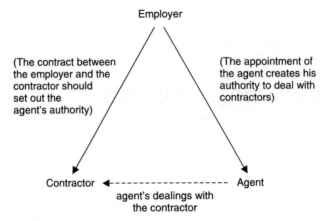

Figure 2.1 The authority of agents

engineer or representative of the employer is the employer's or purchaser's *agent*. Agency is an important legal concept, so it is important that the contract should not only make it clear who that person is, but also: whether or not duties may be delegated; how the engineer is to be replaced, if ever it is necessary; what functions are to be carried out by the engineer; and what limits there are to the authority of the engineer. As far as the engineer is concerned, his authority is contained in his appointment, his allocation to a particular project, and his instructions from his employer. As far as the contractor is concerned, the authority of the engineer is to be found by a careful reading of the contract.

Disputing the decisions of the engineer

The engineer or representative of the employer is, as already stated, the employer's agent. As such, the engineer has a duty of care and skill towards the employer. It may be that the contract will contain a term that the engineer must act fairly, and it may be that in some contracts such a term will be implied. However, this does not alter the fact that the engineer has contractual duties only towards the employer: there is no contractual link between the engineer and the contractor. The conclusion to be drawn is that unless the contractor wishes to be placed in the unhappy position of having to sue the employer, if there is an irreconcilable dispute with the engineer, the contract ought to contain clear and workable procedures for dealing with a dispute between the contractor and the engineer. Such procedures should give as much opportunity as is possible for the parties to resolve the dispute or difference of opinion without the need to call upon the services of an arbitrator, but arbitration may well be needed to be provided for as a final impartial view on an issue in dispute.

In the case of *Pacific Associates Inc. and R. B. Construction Ltd* v. *Baxter and Others* (1988) the Court of Appeal had to consider whether or not a contractor could bring a claim for damages against the engineer, if, as alleged, the engineer acted negligently or failed to act fairly and impartially in administering the contract. The particular claim was on the basis that the engineers should have certified the contractors' claims for increases in the rates of work being done under the contract, which was a contract for dredging and reclamation in the Dubai Creek Lagoon. The court held that the duty of the engineers arose from their contract with the employer, the Ruler of Dubai. They owed a duty of care and skill to the employers and a duty to act fairly. But the engineers owed no duty of care to the contractors, and were liable to the employer alone for any failure to act properly. As far as the contractors were concerned, their rights were to claim against the employer, or to take the matter to arbitration. Accordingly, the claim for £45 million failed.

Could an engineer be liable to the employer for failure to carry out duties with care and skill?

The answer to this question is yes, although such instances do not occur frequently. The engineer has the same duties and liabilities as an agent. If an engineer were to fail to carry out his duties properly – such as design, approval of drawings, approval of materials, or certification of work – the engineer could be liable to the employer either for breach of contract or for negligence. In New Zealand, an architect and a consulting engineer were held to be liable for a faulty design of the structural part of a sports centre, in the case of *Bevan Investments Ltd* v. *Blackhall and Struthers* (1973). In Canada, naval architects were held to be liable for failure to make further inquiries about the suitability of materials to be used in a project. In England, in the case of *Pirelli General Cable Works Ltd* v. *Oscar Faber and Partners* (1983), the same principles of liability were accepted, although the liability in question was not established in this particular case because the statutory time limits, then in force, for bringing a claim had been exceeded. Time limits will be discussed in detail later.

The employer or purchaser

Either of these two expressions is used to describe the person or company or other organization purchasing goods and services from the contractor. It is important that that person, company or organization should be clearly named in the contract, since accuracy of name and address is needed if ever legal disputes or problems of insolvency arise. One question that should be dealt with in the conditions of contract is whether or not the employer or purchaser may assign – that is to say, transfer – legal rights. Duties may not be assigned, so the employer cannot assign the duty to pay. But rights and benefits are assignable unless assignment is prohibited by the terms of the contract.

If the employer or purchaser drafts the conditions of contract, it is unlikely that there will be any restriction on the assignment by the purchaser of rights under the contract. It is not in the purchaser's interests to have such a restriction, since the benefit of an engineering contract is more valuable if rights under it can be transferred. If the conditions are written by the contractor or by an institution purporting to draft neutral conditions, then there may be some form of restriction, such as: '"Purchaser" means the person named as such in the Special Conditions and the legal successors in title to the Purchaser but not (except with the consent of the Contractor) any assignee of the Purchaser.' (I. Mech. E/IEE/ACE.: Model Form 1 [1988]) A clause such as this is intended to prevent any third party purchasing the works from the purchaser from enforcing any rights under the contract against the contractor unless the contractor has consented to the assignment.

The contractor

Similar considerations apply as far as the identification of the contractor is concerned. A contractor may not, by law, assign any of the duties arising under the contract. There is likely to be an equivalent restriction to that already mentioned, as far as the assignment of rights is concerned, but exceptions may be made to this so as to permit the contractor to charge money owed to him by the employer in favour of a bank. A more obvious issue which arises in the case of contractors is the question of whether or not delegation, sub-letting or sub-contracting is permitted, and if so, to what extent.

The employer or purchaser may well, for a number of possible reasons, wish to restrict or to control the contractor's right to sub-contract in some way. The reasons could include: the wish only to have the contractor, who may have been chosen on account of high standards of workmanship, perform the work; or it may be that the employer has relationships with specialist sub-contractors, and would prefer work to be sub-contracted to them; or the reason for control may be in order to exclude sub-contractors whose record is not as good as that required.

The law starts with the view that the contractor is entitled to delegate or sub-contract work, provided that the contractor remains in overall control and does not delegate responsibility. Any limits on the right to sub-contract must therefore be contained in the contract itself. These limits may be phrased in a number of possible ways. For example, certain sub-contractors may be *nominated*, and this procedure clearly prevents the contractor from choosing any other sub-contractor. Alternatively, the contractor may be required to chose certain sub-contractors from a list of *preferred* sub-contractors drawn up by the employer. Alternatively, the contractor may be required to submit his own list of sub-contractors to the Engineer for approval, and may not sub-contract until such consent has been received. Some forms of contract, which contain such restrictions, make certain exceptions, for example, for minor contracts, or purchases of materials by the contractor.

Obligations of the parties

The obligations of the parties, as set out in the contract, together with the price, are the heart of the contract. In this part of the contract will be found the answer to the question of which of the following obligations the contractor is to be responsible for:

- Design
- Manufacture
- Delivery
- Construction
- Commissioning
- Testing
- Correction of defects
- Maintenance
- Maintenance of supply of spares
- Obtaining licences
- Coordinating work of other persons

This should be carefully set out, because there are no particular rules of law about, for example, the testing of equipment. Tests may be standard or special. They may be carried out at the place of manufacture, or after delivery. They may occur before or after the taking over of the equipment by the purchaser. They may be linked to the terms of payment or they may not. It is entirely up to the parties to state their intentions in the contract.

Linkage of terms

The terms of a contract which deal with the obligations of the parties are numerous. However, it is important to bear in mind that they do not work in isolation, and in particular will be linked by reference to documents

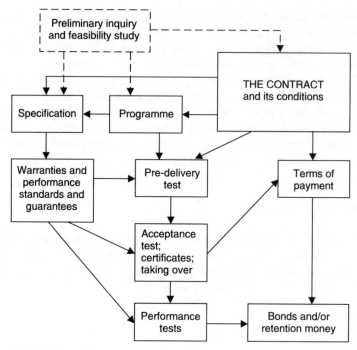

Figure 2.2 *Linkage of terms*

such as a Specification, a Programme, the terms of payment, and probably to a series of agreed tests and certificates. There may also be interaction between these obligations and mechanical guarantees or warranties, or guarantees as to the performance of equipment. These in turn may be backed up by financial guarantees or bonds. Figure 2.2 illustrates an example of the possible linkage of terms of an engineering contract.

Obligations of the purchaser or employer

The contract should set out what these obligations are, not only for the information of the contractor, but also to clarify what might otherwise be legally contentious issues, and to make it clear what are the limits of the employer's duties. Examples of these obligations include the following:

provision of utilities and power;
giving of access to the site;
provision of free-issued goods and materials;
provision of lifting equipment.

However, having given this list purely for the sake of example, this is not to say that all of these will always be employers' obligations. It is a question of what is agreed in each particular contract.

Obligations in which both parties may be involved

There are certain obligations in engineering contracts about which one cannot generalize, and regarding which the contract should, if possible, provide for a number of possibilities. An example of this is the provision of drawings, and the exercise of various functions in connection with drawings.

In general, it is the contractor who has to submit drawings, and for this purpose the engineer usually acts on behalf of the employer in approving or disapproving those drawings. Most engineering contracts are written in such a way that responsibility for drawings, models, samples, etc., remains that of the contractor submitting them, notwithstanding any approval given by the engineer. This is because there have been instances where a contractor has contended that the approval of the engineer has in some way relieved him of liability; the provisions mentioned usually have the effect of making the engineer's approval a necessary requirement for the contract to proceed, without in any way relieving the contractor of his responsibilities under the contract. A well-written contract should also make the time limits for submission of the drawings and other information a part of the agreed Programme, and should state a number of days within which the engineer must either approve or disapprove them, or else be deemed to have approved them.

It is possible that drawings, and certain other information necessary either for the tender or for the contract, may be supplied by the purchaser and/or the engineer to the contractor. The contract should deal with this possibility and in particular should distinguish between errors, omissions and discrepancies on the part of the purchaser or engineer, and similar problems which are caused by the contractor. In each case there may be delays and costs, and one or the other of the parties will have to bear these.

The obligation to submit a Programme

Conditions about time for performance of obligations are an important part of a contract (as will be discussed more fully later on). In some contracts, such as sales of goods, or minor works, the delivery date or date for completion of work is all that is needed by way of a condition about time. With more complex engineering contracts a programme is needed which will show, among other things:

1. a detailed sequence of work, including design, manufacture, delivery, installation, testing and commissioning;
2. times for submission of drawings and other documentation;
3. numbers, at stated times, of contractor's staff expected to be on site;
4. times when utilities and/or equipment to be provided by the purchaser will be required;
5. times when any free-issued goods or materials will be required from the purchaser;
6. times when accesss to the site will be required;
7. times when work done by other contractors, and connected with the contract works, is required to be ready;
8. times when the purchaser must provide information or approvals which are required by the contract.

Variations

A contract consists of a promise to perform specified work, or to carry out certain specified services, or to deliver specified goods, in return for a price or other reward. Whatever has been agreed is fixed and unalterable once the contract has been made, unless the contract provides to the contrary. Thus, a customer normally has no right to change his requirements with regard to the specification or quantity of goods or services, unless there are definite terms permitting this.

However, because of the longer duration and greater complexity of engineering contracts, as well as because of technical problems and the need for certain works to interface with other works, it is recognized that in such contracts there will usually be conditions permitting 'variations', or 'changes', (which bear the same meaning, although the word 'variation' is more common in Britain, and the word 'change' is more common on the other side of the Atlantic). The important thing is to examine the particular condition *before*, as well as after the contract is agreed, since the agreed system of variations can have an effect on the balance of a contract, affecting estimates of time as well as prices.

Broadly speaking, from the point of view of variations, contracts fall into three groups:

1. those contracts that do not permit variations at all;
2. those contracts that do permit variations, but which place limits upon the extent to which variations may be made without the consent of both parties;
3. those contracts which allow for variations and which do not contain any express limits.

If a contract does not provide for variations, then a party wishing to alter some aspect of the works or the programme must place his proposal before the other party and obtain agreement. This, in effect, is an *amendment* of an existing contract, and should be signed by both parties. The contract which allows for variations simply short-cuts this rather lengthy process, by having conditions which give the *prior consent* of the contractor to variations required by the purchaser or the engineer. As a consequence, the parties need only follow the agreed procedures, and no amendment of the contract is required. Moreover, because prior consent has been given, the purchaser will, in most cases, find it unnecessary to go through the bargaining process that would be needed to obtain an amendment to the contract. In standard engineering contracts, the engineer is the agent of the purchaser for the purposes of ordering variations, and so that he does not exceed his authority, specific procedures are usually stated. The contractor should pay particular attention to these procedures, especially those about payment and allowance of time for variations.

Limits to the power to order variations

These sometimes exist in engineering contracts, although they are by no means standard. If there are no express limits in the conditions of contract, then the law will imply limits at a certain point, but reliance upon implied limits is risky and contentious. For this reason, employers and contractors have often found that 'express limits' are advisable. Express limits protect the contractor, in so far as the contractor can, at a certain point, object to the extent of the variations proposed and negotiate new terms; this would be of importance where, for example, a variation clause tied the contractor to specific rates or prices. The employer or purchaser, too, can benefit from such conditions of contract, since the employer using such a clause will have the security of knowing that the engineer cannot place variation orders which will raise the overall price of works beyond an agreed limit without the matter having to be referred back to the employer for approval. Figure 2.3 shows the structure of limits that may be placed upon variations.

What amounts to a variation?

A variation is a change in the nature or specification or manner of performance of work to be done under the contract. It can include additions or omissions. Often its meaning is defined by the contract. However, points of law occasionally arise as to what is or is not capable of being a variation.

28 Structuring contracts

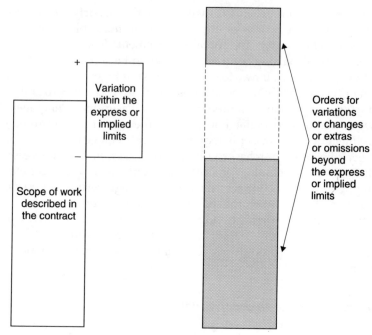

Figure 2.3 Limits to variations which may be made in commercial and engineering contracts

Instructions to do what the contractor is already bound to do

These normally do not count as variations. They are simply day-to-day supervision by the engineer, or else they are *clarifications* of the contractor's existing obligations. Occasionally, an issue has arisen because such an instruction has been documented as a variation and later on it is realized that no variation is involved. Sometimes there has been an agreement to pay additional sums. Can such agreements be retracted? The answer in most cases is yes, since the contract has already allocated a price to the existing work. There is therefore *no consideration* for a promise to pay extra.

However, in the case of *Williams* v. *Roffey Bros & Nicholls (Contractors) Ltd* (1990), the Court of Appeal had to consider an unusual situation, which may have cast the law in a new light. In this case, a sub-contractor sued the main contractor for additional payments, over and above the sum of £20 000 which had been agreed as the contract price for the sub-contracted work. During the course of the work, the sub-contractor ran into difficulties, caused partly by problems of supervision and partly by having underpriced the work in order to be awarded the contract. The sub-contractor approached the main contractor and requested additional money, which was agreed by the main contractor, who was concerned to avoid liability for delays to the main contract. The sub-contractor carried on work, but did not receive the extra money promised. The sub-contractor then stopped work and sued. The court held that although the work to be performed by the sub-contractor was work that the sub-contractor already had to do, there was a form of consideration for the promise to pay extra, and accordingly the extra payment was binding on the main contractor. The consideration was that there was some doubt as to whether it was possible for the sub-contractor to complete his work, and in return for extra payment the sub-contractor had given the main contractor the benefit of a promise that the work would in fact be done on time. This benefit was brought about without any threats or duress on the part of the sub-contractor, and consequently was valid consideration.

Variations which are made without authority

All variations must either be given or confirmed by the employer, or else must come from the authorized person. This will usually be the engineer or other representative of the employer. If a variation does not have the required authorization, it should not be carried out by the contractor, for two reasons. One is that there would be no duty on the part of the employer to pay for it. The other is that it could constitute a breach of contract by the contractor, who would, without proper instructions, be substituting a different thing for the work required to be done under the contract.

Variations which lack the required formality

Most conditions of contract provide for variations to be confirmed in writing, and this is good practice in any case, as there are many particulars to be recorded about variations and this makes it unwise to give them orally. If there are no conditions in the contract which impose a requirement of writing, then in theory an authorized person may give an oral variation instruction, and this would be contractually binding in just the same way as the making of an oral contract in the first place. If the contract states that variations must be in writing, this makes an oral variation order invalid unless and until confirmed in writing. It is possible that the employer (as compared with the engineer) might *waive* the requirement of writing, for example in an emergency, but this would be an unusual occurrence, and in most cases both parties should steer well clear of oral variations. Figure 2.4 is an example of a form of variation order.

```
DATE ................        CONTRACT NO: ................................................
                             DESCRIPTION OF WORKS ..................................
                             ...............................................................................
                             VARIATION ORDER NO: ....................................
TO ................................................................................ (Contractor)
You are directed to carry out the following variations to the Works:




Authorized by ............................................. (ENGINEER)

Authorized by ............................................. (EMPLOYER) (Where applicable)

This variation results in the following adjustment to the
Contract Price:

CONTRACT PRICE: ..............................................................................
NET INCREASE/DECREASE
in the Contract Price
resulting from previous
variation orders ..................................................................................
NET INCREASE/DECREASE
resulting from this
variation order ...................................................................................
TOTAL SUM NOW DUE TO
CONTRACTOR .....................................................................................

Time for Completion to be adjusted as follows:
NET INCREASE/DECREASE resulting from this order ..............................days
TIME FOR COMPLETION PRIOR TO THIS ORDER .............................................
TIME FOR COMPLETION AS ADJUSTED BY THIS ORDER ..................................
```

Figure 2.4 *An example of a variation order*

Waiver

Waiver is a legal concept which is in many ways similar to variation. However, in engineering contracts it is generally recognized that there is a difference between the two things. A variation is something which is either bilaterally agreed by the two parties, or else which is expressly provided for by the contract. It is usually structured, and procedures of the kind already mentioned are usually laid down. As has been seen, specific variation order forms are quite normal in engineering contracts. Waiver has one thing in common with variation, and that is that there is an alteration of something that was provided for in the contract, and the parties accept that alteration. However, waiver differs from variation in that it is the exception rather than the rule. An employer does not request a waiver, unless exceptional circumstances arise.

For example, an employer who has agreed to take delivery of goods on a certain date may be unable to do so for unexpected reasons. If he asks the contractor to agree to deliver the goods at a later date, he is, in effect, asking for a relaxation of the terms of the contract. This relaxation or concession is called a waiver. Unlike a variation, there need be nothing offered in return: no consideration is required for a waiver to be binding, whereas a variation under which the contractor agreed to provide something extra would only be binding if there were some consideration, such as additional payment, in return. Any concession or relaxation of the contract, whether by the employer or by the contractor, can amount to a waiver. The test of what is a waiver is one of commercial intention of the party making it, but once a waiver has been made, the party making it cannot go back on it. For this reason it is quite common in engineering contracts to see conditions which state that any delay or forbearance by the employer in exercising his rights is not to be construed as a waiver.

Tests, taking-over, acceptance and rejection

Engineering contracts usually provide for tests, and there are no rules of law as to when those tests are to be carried out, or at what place, or in whose presence; nor are there any rules as to the type of tests to be carried out. It is entirely up to the parties to make the necessary provisions in their contract. The requirement as to tests, and the place and stage of the programme at which these should take place, should be stated in the 'Conditions' of the contract. The technical details, data, standards and performance criteria should be set out in schedules or annexes to the contract, which should be formally designated as contract documents and should be referred to in the Conditions. The reason for this is that different types of tests have different consequences, particularly as to payment and acceptance, and the distinction should be clearly made in the contract.

Tests or inspection before delivery

These should be provided for in the contract if the purchaser requires such tests or inspections: normally a purchaser of goods or services has no rights of testing or inspection prior to delivery. This is because such matters necessarily have disruptive effects and possibly have cost implications, and therefore have to be expressly agreed upon by the parties if the purchaser is to have any such rights. Usually the provisions in the Conditions will be of a fairly general nature, permitting the engineer to inspect or test at 'reasonable times' to be agreed upon by the contractor and the engineer. These times may then, if the parties wish, be set down in the Programme, or they may be left to be arranged by the giving of notice.

Usually, inspection and tests prior to delivery are for the information of the purchaser or engineer. Their purpose is to avoid misunderstanding and any delays which would be caused by delivery of incorrect or defective goods to site. They do not mean that the purchaser or engineer has accepted the goods, since there may be further work to be done or services to be performed on site, and further tests to be carried out. The wording of conditions about inspection and testing prior to delivery usually makes this clear, and any document or certificate issued has to be read in the light of such conditions. The contract Conditions may also make arrangements about costs and expenses, particularly if expensive labour and apparatus is needed, and particularly if it has to be repeated as a result of failure of a test.

Tests on completion

These are perhaps the most important tests in an engineering contract, because they would be needed even if the inspections or tests prior to delivery were treated as optional. The date for completion of works is, of course, set out in the contract, but it is usual to provide for the giving of notice by the contractor to the engineer that the tests on completion are to take place. The reason for this is that completion may take place earlier than the date stated in the contract, or it may take place at a later date. The significance of the tests on completion is not only that a major part of the payment of the price may depend upon whether or not these tests are passed, but also the fact that after a successful series of tests on completion the works will, in a legal as well as an engineering sense of the word, be deemed to be complete. This means that the purchaser may take them over and put them to use. It also provides the benchmark for the precise date on which the works are completed for the purposes of calculating whether or not completion was on time or delayed. If the works are completed – in this sense – on or within the date for completion, then the contractor can have no liability for delay, even though subsequent events may cause delay in the effective operation of the equipment. If the works are not completed on time, then the contractor may incur liability under the conditions of contract for delay. Exactly what this liability is, and how it is calculated, will be discussed later.

As with other series of tests, the details of the tests on completion should be set out in the Conditions and in schedules. In particular, the consequences of passing or failing the tests on completion need to be made clear. Failure may, for example, entail rejection and repeating of tests; it may, ultimately, entail the termination of the contract if repeated tests are failed, and if stated time limits are exceeded. The successful carrying out of the tests will usually entail the entitlement of the contractor to a certificate, and a payment of a significant proportion of the

---

TO: .... CONTRACTOR

NAME OF PURCHASER:

DESCRIPTION OF WORKS:

1. It is hereby certified in accordance with Clause...... of the General Conditions of Contract that the Works specified in the Schedule below were completed (subject to the exceptions noted below which do not affect their commercial use) and have passed the Tests on Completion. The Purchaser is deemed to have taken over the said Works with effect from ....(*date*)

   SCHEDULE OF WORKS ....
   ....

2. A list of items which remain to be completed is annexed to this Certificate of Completion, and the Contractor is required to complete them within ....days of the date of this Certificate.

3. The Warranty period (Clause ....of the General Conditions of Contract) commenced on ....(the date of Taking Over).

4. The sum of £ ....due to the Contractor on the issue of the Taking-Over Certifcate shall be paid by the Purchaser to the Contractor within................days of the date of this Certificate.

5. The Date for Completion, having regard to claims for extension of time received prior to the date of this Certificate, expired on...

SIGNED: .... *ENGINEER*

Figure 2.5 *An example of a taking-over certificate*

34 Structuring contracts

price. The title of the certificate will depend upon the terms of the contract, but commonly it is called a 'taking-over certificate', an example of which is given in Figure 2.5.

Performance tests

These may be provided for in a contract, although they are not as essential in a legal sense as the tests on completion. They may be described as a series of tests which are to take place during the warranty or 'defects liability period', after the taking over of the works by the purchaser. The purchaser will often be using the works commercially during the period of time during which the performance tests are to take place, and will often have paid the greater part of the price by this time. The aim of performance tests is to ascertain the capabilities of the works or equipment in varying circumstances and under different conditions, and often using different input or 'feed' materials. By doing this the purchaser will gain the benefit of being able to detect deficiencies or defects during the warranty period, and will be able to have them corrected. The purchaser will also be able to ensure, by a series of tests rather than a once-and-for-all test, that the works are capable of the levels of performance provided for in the contract. As with the other types of tests, the details of the tests should be stated in schedules. Unlike the tests on completion, there will be no liability for delay in completion if the performance tests are unsatisfactory, since 'completion' will have taken place before the performance tests have begun. It will therefore be appreciated that it is a matter of fine judgment and technical expertise to decide at the outset, before the contract is made, what matters will be tested on completion and what will be left over for performance tests.

Failure to pass performance tests

If the works with regard to an engineering contract fail to pass the performance tests this usually means that certain agreed standards of performance, or agreed values of output or consumption figures, have not been attained: for example, a machine for the production of metal components may have been installed on the basis that it can produce 10 000 units per hour; or it may have been installed on the basis that its fuel consumption will be x units per hour; plant for purification of water, fluids or gasses may have been installed on the basis of a level of purity that can be attained; or an item for use in the air or in space may have been delivered on the basis that its weight will not exceed a certain figure. In any of these cases, performance tests will ascertain whether the facts and figures are true and will remain true in certain controlled conditions. The tests on completion will have already informed the engineer that the item has been completed and is free from defects and is capable of performing as required. The performance tests are to ensure that this can be verified over a longer period and in a wider range of circumstances.

If the works or equipment do not perform as required the contract will normally provide for modification or adjustment to the works or equipment, and for subsequent re-testing. This, in most cases, is sufficient to cope with such problems. However, all contracts have to foresee worse situations, such as the inability to adjust the works so as to attain the required levels of performance. A good contract should give the contractor a fair opportunity to put matters right in such circumstances, and then should give the purchaser a number of options, so that a commercially sensible choice can be made. One choice might be that of deducting a sum from the purchase price by way of 'liquidated damages', so as to compensate the purchaser for the deficiencies. This only applies if the liquidated damages have been agreed in advance as terms of the contract. It is only worthwhile to do this if the correct compensation for deficiencies can be calculated in mathematical and predictable terms. If it is, for example, a matter of output, this can probably be done. On the other hand, with matters such as weight or purity, it may not be possible to use a scale of liquidated damages, because it may be that the consequences of failure are too deep-seated to be put right by a sliding scale of money.

An alternative would be for the purchaser to have the option of either accepting the works or rejecting them, after a failure to meet required performance limits, and to provide that if the purchaser accepts in such circumstances the price is to be reduced by either an agreed sum, or by an amount to be determined by an arbitrator. It will be noted that if such a provision exists, the purchaser will be entitled to reject, notwithstanding that earlier there had been a certificate of taking-over.

It follows from this that 'acceptance' with regard to completed works in an engineering contract is a difficult expression to define, and can only be understood when the contract has been appraised as a whole, taking into account the series of different types of tests, and the purchaser's rights of rejection. In a simple sale of goods, acceptance means the moment when the buyer is deemed to be unable to reject or to return goods, even if the goods are defective. In an engineering contract acceptance may be defined as the time when, after all tests have been carried out, the purchaser has either certified that the works have passed all tests, or else has elected finally to keep the works, even though they may have failed some of the tests. Often in engineering contracts a 'final certificate' is issued, and this will be evidence that the works have been accepted.

Some legal questions answered

Is a clause in a contract prohibiting the assignment of rights a valid clause?

This has been the subject of some doubt until recently. In the case of *Helstan Securities* v. *Hertfordshire County Council* (1978) a clause in the contract stated that the contractor was prohibited from assigning the contract or any benefit or interest therein. The court held that an assignment of money due, without consent, was invalid as a result of this clause. Nevertheless, there remained some doubt as to what these clauses really meant, and the limits to which they could go. The new test cases on this subject have now gone a long way towards settling the law. The cases are: *Linden Garden Trust Ltd* v. *Lenesta Sludge Disposals Ltd* (1994), and *St Martins Property Corp. Ltd* v. *Sir Robert McAlpine Ltd* (1994). Both cases were concerned with a clause which stated that: 'The employer shall not without written consent of the contractor assign this contract.' The first case was an asbestos removal contract, and the second was a building contract. In both cases the employers assigned to other parties their interests in the properties concerned, together with all rights and the benefits of all contracts. The parties who had purchased these interests wanted to claim damages for defective work against the contractors. The House of Lords held that the prohibitions on assignment were valid, and the assignees were unable to bring claims in their own right against the contractors.

When does the engineer become responsible for errors in drawings and other information?

Mere approval of drawings and information submitted by the contractor does not normally commit the engineer or the employer in any way. This is because most contracts place design responsibility on the contractor, and state that approval by the engineer of drawings will not relieve the contractor of any responsibilities under the contract. The position is different if drawings and information are provided by the employer or the engineer. Then they will have to accept responsibility for errors in that information. The matter becomes rather more complex if the engineer receives drawings from the contractor and suggests or requires amendments to them. What is vital at this juncture is to make it clear where the responsibility lies. For example, the engineer could suggest amendments but make it clear that responsibility for incorporating them into the design, as well as responsibility for the design as a whole, remains with the contractor. Failure to do this can have adverse effects. In the Australian case of *Cable Ltd* v. *Hutcherson Ltd* (1969) the *tender* included design, supply and installation of a bulk storage and handling plant. The contractor submitted drawings and the engineer required amendments to them. The tender was accepted and a formal contract was made which incorporated the amended drawings. However, far from placing design responsibility on the contractor, the Conditions of Contract only stated that the contractor was to 'execute and complete the work shown on the contract drawings and described by or referred to in the specification and conditions'. Towards the end of the work it became apparent that the design would result in subsidence. The High Court of Australia held that because of the wording of the contract, and because of the changes to the drawings required by the engineer, the contractor was not responsible for the adequacy of the design to do the intended work or to take the intended load: the contract was only to execute the precise work described in the drawings, and whether it functioned correctly or not was not the contractor's responsibility.

If a clause permitting variations has no express limits, can limits be implied?

Yes, limits can and have been implied in certain cases. Firstly, the law of agency will always apply to variation orders, so that they will only be valid and binding on the employer if given by the engineer acting within his powers, or another person authorized by the contract. Secondly, a variation must be, by necessary implication, a variation of the work specified. It cannot be work or goods which are or ought to be the subject of a separate contract. One cannot, by use of a variation order or a series of such orders, turn a contract to build a surface ship into one to build a submarine. Nor can one use variation orders to achieve a cancellation of an order (which might seem theoretically possible by a series of cuts), since cancellation and variation are two separate issues. What is more difficult to assess is a case where the difference between a variation which falls within implied limits, and a variation which falls outside those limits is marginal. Implied limits are necessarily vague, precisely because they have not been specified, and a contractor wishing to be able to enforce limits would do well to state those limits in the contract. A commonly used method of doing this is to have a clause stating that no variation when taken together with other variations may, without the consent of the parties, increase or decrease the contract price by more than a stated percentage. This is not an entirely satisfactory method of putting limits on variations, because the cumulative effect of a large number of variations may still fall within the stated price limits, while at the same time imposing considerable strains on the resources of the contractor. However, it is not easy to put an alternative formula into practice.

Two cases may be mentioned to show when and how limits on the power to order variations may be implied by the courts. The first of these is *Parkinson & Co. Ltd* v. *Commissioners of Works* (1950). In this case the contract was such that the amount of profit available to the contractor could not, in any circumstances, exceed £300 000. Variations were ordered which greatly magnified the contract value, which was originally £5 000 000. But the profit still could not exceed the stated sum, so that the contractor could, it appeared at first sight, be required to do an infinite amount of additional work at cost price but without additional profit. In the circumstances the Court of Appeal implied limits so that additions and extras could not *materially* exceed £5 000 000. Any additions or extras would therefore have to be the subject of fresh negotiations.

In the second such case, *Cana Construction Co. Ltd* v. *The Queen* (1973) there had been a tender for the construction of a postal terminal in Canada. The tender was to include the supply and installation of mail handling equipment, which was to be done through a sub-contractor, so the tender included a figure for supervision and overheads and profit. The tender figure was on the basis that the sum for the supply and installation of the handling equipment would be approximately $1 150 000. In the event there was a 'change order' which required the equipment to be of a specification for which the sum required by the sub-contractor would be in excess of $2 000 000. The main contractor asked for a change in terms of payment, which was refused by the employer. The Supreme Court of Canada held that the change order went beyond what was contemplated by the parties, and a terms would be implied that this fell outside the definition of a 'change', and the employer would not be entitled to call for the alteration to the specification without being prepared to agree a different sum in respect of overheads and profit.

3 Price and payment

One person's pay increase is another's price increase
                                                            Harold Wilson

Price

All engineering contracts should contain terms about prices and payment. The two types of terms can be distinguished in the following way: terms about price tell us the sum that the contractor is to receive, or the criteria for reckoning the amount of money that is due. Terms about payment tell us about the timing of the payment or series of payments, and about how payment is to be applied for, and about the method by which the money is to be transferred from the employer to the contractor.

The requirement of certainty

A price does not have to be a specific figure or sum, but it does have to have an element of certainty. One of the reasons why a letter of intent is usually not a contract, is that the terms are still under negotiation, so it cannot be said with certainty what they will be. Terms about price are so important that a commercial transaction may fail for lack of certainty if there is not an agreed, binding, certain method of reckoning the price. Any of the following methods may be used:

1. a fixed lump sum. (Sometimes the word 'firm' may be used instead of the word 'fixed'.)
2. a price calculated on a 'cost plus' basis – that is, the actual cost of an item plus an allowance for profit;
3. a price to be calculated by agreed rates or bills of quantities, and measurement of work;
4. a fixed price, but with re-imbursement of certain expenses;

5 a fixed price which includes 'provisional sums', which may or may not be spent, or which may only be spent in part;
6 a fixed price to include sums for prime cost items, to be spent as directed by the engineer. (The contract may, or may not, as the case may be, provide for an allowance for profit on these prime cost items.);
7 reasonable remuneration for goods provided or for services rendered.

Obviously, an agreement to pay reasonable remuneration is far less certain than a fixed lump sum, but it will not fail, from a purely legal point of view, on the grounds of uncertainty, if that is what the parties have agreed. What *will* fail for uncertainty is a position where it is impossible to say with any accuracy what the parties have agreed, or indeed whether they are in agreement at all.

In the case of *Courtney and Fairbairn Ltd* v. *Tolaini Bros Hotels Ltd* (1975), it was held by the Court of Appeal that an agreement to 'negotiate fair and reasonable contract sums ... based upon agreed estimates of the net cost of the work and general overheads with a margin for profit of 5%', did not amount to a binding and enforceable contract. The estimates mentioned had not been agreed, but were still to be the subject of further negotiations. Lord Denning said:

> Now the price in a building contract is of fundamental importance. It is so essential a term that there is no contract unless the price is agreed or there is an agreed method of ascertaining it, not dependent on the negotiations of the two parties themselves.

It is not the lack of a figure that is fatal to the existence of a contract; it is the absence of a definite, non-negotiable method of ascertaining the price. If, during the negotiations the parties fail to agree the price or the basis for ascertaining the price, there is no contract. However, if the parties each put forward different versions of the price, as happened in a case we have already noted (*Butler Machine Tool Co. Ltd* v. *Ex-Cell-O Ltd* (1978)), there will be a contract when one party accepts the other's proposal along with the other terms offered.

Prices ruling

In some commercial contracts, usually for materials or for labour and materials, or for smaller items, one may sometimes see in the seller's conditions of sale a condition that the price payable will be the 'price ruling' at a specified date – such as the date of delivery. The *Butler Machine Tool* case was about just such a condition, which was in the seller's conditions of sale. In the event, it was held that the buyer's conditions applied, which stated that the price was fixed and not subject to increases. We may assume that as this was an ordinary commercial contract between two businesses, the clause in the seller's conditions was a perfectly legitimate way of calculating prices, and would have been contractually valid if the buyer had accepted it. There is, perhaps a problem of interpreting such a clause, but we may again assume that a court would take it to mean the price of those goods, materials or services which the seller applied to customers in general at the relevant time. Evidence of this would be a list price. If there was no evidence of a price ruling which was any different from the quoted price, then the quoted price probably would be the price ruling. It should be noted that if the buyer is a private consumer, as compared with a business, such a condition may be invalid under the EC Directive on Unfair Terms in Consumer Contracts. This Directive, which will be discussed in more detail later, provides that terms in contracts made with consumers may be regarded as unfair in certain circumstances. The Annex to the Directive gives examples of terms which may be regarded as unfair, and one of these examples is a term:

> providing for the price of goods to be determined at the time of delivery or allowing a seller of goods or a supplier of services to increase their price without in both cases giving the consumer the corresponding right to cancel the contract if the final price is too high in relation to the price agreed when the contract was concluded.

What is included in the price?

The price in an engineering contract usually includes everything that is required to achieve the performance of the works to be carried out. But although this could be deduced by implication from the contract terms and specification, it is better if the details of what is or is not included are expressly stated in the contract. Matters such as the supply of spares, tools, drawings or manuals, or the provision of additional services, or the provision of training, are capable of causing differences of opinion, and clarity on these matters is much to be desired. Estimators have to take into account all matters which may affect the price, and they should bear in mind that if an error in understanding what is required results in the underpricing of a tender or other offer, this will not invalidate the offer. If the other party accepts the offer in good faith, then it is a contract. Normally, the only time that a contract will be rectified on the ground of an error affecting the price is if the two parties have already reached agreement, and subsequently the document setting out the agreement alters the terms, and this is known to one party but unknown to the other. So in the case of *Roberts & Co.*

Ltd v. *Leicestershire County Council* (1961), a tender for works stated that the period for the completion of the works was eighteen months. However, the contract was drawn up by the other party, and it gave a period of thirty months for the completion of the work. This alteration had not been notified or pointed out to the tenderer, and if it had been, would have resulted in a higher price. The court, on the application of the tenderer, rectified the contract and substituted the shorter period for the longer one.

By way of contrast, in the case of *Ibmac Ltd* v. *Marshall Ltd* (1968), the contractors had quoted a price for the building of a road. The quotation was accepted and the work started. The site was at the bottom of a steep hill and the contractors had not foreseen difficulties regarding surface water. This created problems which would cost additional money to put right, and the contractors expected the employer to pay for this. However, it was held that under the terms of the contract it was the responsibility of the contractors to complete the work at the price quoted and accepted. It was not the responsibility of the employer to put the site into a state which would make it easier or cheaper to do the work. It follows from this case that those who enter into contracts should, before quoting or tendering make a careful assessment of all conditions which may affect the price and timing and feasibility of the work, and take full account of them in the tender or quotation. Many forms of engineering contract have a clause with a heading such as 'Basis of Contract Price', and this clause will often make it clear that it is for the contractor to inform himself fully about the condition of, and circumstances affecting, the site.

Payment

The terms of engineering contracts vary a great deal as to terms of payment. Attempts may be made to standardize conditions of payment, but the parties will almost certainly vary them during the course of the negotiations, so there is little point in attempting to set out any such terms as standard in this work. The points of difference from contract to contract are:

1 whether there is to be a single payment or a series of stage payments;
2 if there are to be stage payments, the 'milestones' for each stage payment, and the proportion of the price allocated to each stage (which will vary a great deal);
3 the currency of payment (which is entirely for the parties to decide upon);

4 The method of payment (for example, whether it is to be by cheque or bank draft or by letter of credit);
5 how each payment is to be secured in favour of the employer, if at all (security may be by way of bond or guarantee, or by the passing of property in goods).

Stage payments

If a contract, whether it is on a lump sum or a cost-plus basis, or even on a measurement basis, is one where the contractor has to achieve complete performance of the work (other than the warranty), before any payment is due – lawyers call this an 'entire' contract. Entire contracts suffer from a problem from the points of view of both parties. The contractor is at risk, as there is a possibility that for some reason complete performance may not be achieved, in which case no payment will be due. With an entire contract, there is no room for the contractor to claim a proportion of the payment that would have been due, when the work is only partially completed. So in *Ibmac Ltd* v. *Marshall Ltd*, the facts of which have already been considered, the contractor abandoned the project, as there was a difference of opinion over who should pay the additional cost of putting the site in order. The contractor claimed a proportion of the money that would have been due on completion. It was held that no money was due, and on this point the Court of Appeal reversed the original decision of the trial judge, who had awarded a proportion of the price to the contractor. In this particular case, there was some doubt from the wording of the contract as to whether it was a lump sum contract or a contract under which the price was to depend upon measurement and the application of bills of quantities. The Court of Appeal held that this point did not affect the outcome, since, in any case, it was still an entire contract.

From the point of view of the employer, one of the defects of an entire contract is that there may be less scope for monitoring the progress and achievement of the contractor. It cannot be said for certain that distinctions of this kind between entire contracts and contracts providing for stage payments can always be made, since everything will depend upon the monitoring procedures. However, a contract under which the contractor is to be paid for reaching certain identified milestones of achievement is more likely to give the employer opportunities for monitoring progress than is an entire contract. This may be particularly true of contracts for research and development.

Stage payments may, according to the terms of the contract, involve virtually any number of stages. The most typical example of stage payments is the kind that breaks down into three stages: firstly, a stated percentage of the price is payable when the contract is made; secondly, a

44 Price and payment

major percentage of the price is payable when an item is either delivered or installed and handed over to the employer; thirdly, a smaller percentage is paid some months after the taking over of the equipment or works. Having said this, it will be clear that the number of stages could be considerably more if there are more activities to be carried out, or if there are several phases of work. The key to the understanding of the terms of payment is to identify what has to be done to earn payment and to gain the release of money. In many cases a Certificate will need to be presented to the employer, and it is important to know when and how and by whom the Certificate is to be issued.

Retention money

The concept of retention money applies in building and in engineering contracts, as well as some other types of commercial contracts. The basic idea is one of security for performance: the employer or purchaser 'retains' some of the price until a stated date or period after delivery or completion has elapsed. This often, but not necessarily, coincides with the defects liability or warranty period. Typical retention sums may be five per cent, or ten per cent, or sometimes even a higher percentage of the price. At least two different ways of creating a retention fund may be identified. Which way applies in any given case will always depend upon the terms of the contract. Without express terms of contract conferring the right to hold back money as retention monies, there is in fact no such right in favour of the employer, and all of the contract price must be released at the latest on the date of completion, if not earlier, according to the stages of payment agreed.

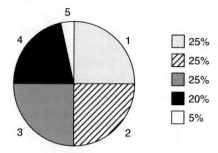

Note: In this illustration the final stage of payment is the 5 per cent of the price which is outstanding at the end of the contract. If it is retained for any time the question is how to secure the retention fund.

Figure 3.1 *Retention money*

Price and payment 45

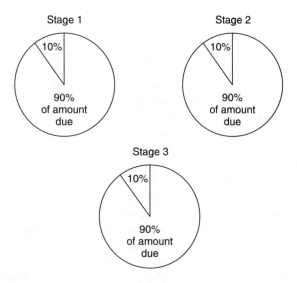

Notes: In this example, 10 per cent of each stage payment is held back, until the agreed date for release of the retention fund.

Figure 3.2 *Alternative method of deduction of retention money and the creation of a retention fund*

One way in which a retention fund may be created is simply by providing for stage payments, and by arranging that the last stage payment will be retained by the employer, after the completion and taking over of the works, until a specific period of time has elapsed. The period of time may be a few months, or as long as two or three years, again depending upon what has been agreed. This method is illustrated by Figure 3.1.

An alternative and rather more complicated way of creating a retention fund is for the employer to agree stage payments with the contractor, and also to agree that as each stage payment is due, a fixed percentage of the stage payment that is due will be held back as retention money. This fixed percentage may, for example, be 5 per cent of each amount due as a stage payment. Again, it will be kept by the employer until a specified period of time after taking over has elapsed. This is illustrated in Figure 3.2.

Legal considerations regarding retention money

Any delay in making payment in full for goods or services received can only be a benefit to the customer, who is obtaining an element of credit as well as the opportunity to detect any defects, which may then be used as a ground for non-payment of the retention money. The benefits to the customer are an equal and opposite detriment or risk to the contractor: the contractor suffers a loss of cash flow, as well as the risk (which is by no means always theoretical) that the customer will find some reason to refuse to pay the retention money. The most serious risk of all is that the customer may become insolvent while holding large sums of retention money owed to the contractor. Contractors should be aware that this money, even if legally due to them, is still an *unsecured debt*, in the majority of cases. The meaning of unsecured debt will be considered in more detail later, but for the time being the point may be illustrated by the following case: *MacJordan Construction Ltd* v. *Brookmount Erostin Ltd* (1991). In this case a builder agreed terms of contract under which the employer, a property developer, kept back a sum of money as retention money. Subsequently, the property developer borrowed money from a bank and granted the bank a security by way of a floating charge over its assets. The property developer then became insolvent and the bank appointed administrative receivers. The contractor claimed that its right to the retention money took precedence over the charge created in favour of the bank, but the Court of Appeal held that this was not the case. No valid charge or trust had been created in favour of the contractor, and consequently the contractor's right to the retention fund was unsecured, whereas the bank was a secured creditor.

Performance bonds or guarantees

Due to the drawbacks of retention money, from a contractor's point of view, a contractor may wish to negotiate an alternative form of security for the employer which may serve wholly or partially as a substitute for retention money. This is the performance band or guarantee, which is a financial guarantee for due performance of the contract, and which is given by a bank or insurance company or other surety. The employer holding a performance bond or guarantee has a form of security against defective performance by the contractor. It is, from the employer's point of view, less convenient than the simple remedy of keeping retention money in hand, but from the contractor's point of view it has advantages: the contractor gains a superior cash flow and, depending upon the form and wording of the bond, there may be less risk in the event of the insolvency of the employer. In some engineering contracts, the employer requires *both* a bond, and retention money, and the cumulative effect of these must be kept in mind.

Price and payment 47

Times and dates for payment

Engineering contracts may provide that one or more invoices must be presented by the contractor, in order that payment shall be made. These may be periodic invoices, such as monthly invoices, or they may be invoices rendered as soon as delivery or completion or some other event has occurred. It is unlikely that payment will be made as soon as an invoice is submitted or sent, and it is likely that the contract will provide for a period, such as thirty days from receipt of the invoice, within which payment is to be made. Another common payment procedure is to provide that payment will be made at the end of the month following the month of receipt of the invoice. This is popularly known as 'net monthly account' terms of payment, but the expression is not recognized by law, and should not be used unless it is defined in the contract or otherwise agreed by the parties. If no period for payment is specified in the contract, then common law principles apply and these are that payment is due on delivery of goods or on completion of work. If the contract is an 'entire' contract, then 'delivery' means delivery of all the goods specified, and 'completion' means completion of all the work specified. Strictly speaking, unless a period such as thirty days from the date of the invoice is agreed, then payment is due *immediately* the events giving rise to payment have occurred. Thus, for example, unless otherwise stated, all sales of goods are on a cash on delivery basis. In practice this principle is almost invariably overidden by the terms of the contract, or by other arrangements between the parties, such as account facilities or payment by letter of credit.

Some legal questions answered

Is the contractor entitled to charge additional amounts if the costs of labour or materials or transport increase?

The answer to this question depends upon the express and implied terms of each contract. In general, a price is presumed to be a fixed price, and not subject to increases, unless there are indications to the contrary in the contract. In some forms of engineering contract there are clauses about the price which contain different options for the parties to agree upon at the time of negotiating the contract depending upon whether or not they are prepared to provide for any part of the price to increase or decrease in line with external costs.

Is interest payable on sums due under the contract?

Some forms of contract specifically provide that the buyer, or employer, will be liable to pay interest on sums owed to the contractor which are overdue. If such clauses exist in contracts, they are usually enforceable, provided that they are not unfair contract terms or penalties. In short, interest rates which fall within normal commercial rates are a valid and legitimate device to offset the risk and inconvenience of late payment. Many sellers and contractors include interest clauses in their own conditions of contract, and several national standard forms of engineering contract do so as well. It is far less common to find such clauses in conditions of contract written by purchasers, although examples do exist.

Without an express term providing for interest, the question arises as to whether or not there is any general or statutory provision of law under which interest can be claimed in respect of sums paid late or overdue. A great deal depends upon the law applicable to the contract. Under several systems of law, mandatory interest arises either as soon as payment is overdue, or a short time afterwards. However, under English law there is no such rule, although at least one abortive attempt has been made to put such a law on the statute book, and various government departments have held extensive consultations on the subject. If payment becomes overdue to the extent that the debt becomes the subject of court proceedings, then the courts have power to award interest. However, the reality is that where payment is overdue, settlement is usually made before court proceedings arise, and in such circumstances, unless the contract provides for interest, no interest is payable.

Can a contractor refuse to deliver goods on the ground of failure by the purchaser to pay for previous deliveries?

This is a more complicated question than appears at first sight. As always, one should first look to the express terms of the contract(s) in issue, since such contracts may well contain specific provisions dealing with non-payment. Under the law of sale of goods, a seller does have a *lien*, that is, a right to withhold deliveries, unless credit has been given, if the goods have not been paid for. However, this does not give the seller the right to withhold goods to be delivered under a later contract on the grounds that goods delivered under an earlier contract have not been paid for. Only the most specific terms in a contract could have that effect. The concept of *lien* is, however, only relevant to sale of goods contracts: it is seldom applicable, as such, to engineering and construction contracts. This is because of the more complicated and extensive framework of rights and duties within engineering contracts: the contractor is under an obligation to meet specific programme and performance requirements, and it will normally be implied that the law of *lien* is overridden by these obligations.

When does failure by an employer to make payment to the contractor entitle the contractor to stop work?

Under common law, time for payment is a term of the contract, but time for payment is *not of the essence*. What this means is that although a failure on the part of the employer to make payment when due is a breach of contract, it is not the kind of breach which gives rise to immediate consequences: the contractor wishing to take action must first give notice to the employer. Under common law this period of notice must be reasonable so as to allow the employer the chance to put matters right: the law has long recognized that in engineering contracts it would have most unfortunate consequences for the employer if such a contract were liable to be terminated for every failure to pay, no matter how small, no matter how easily it is put right. In most engineering contracts, so as to be certain about these matters, a period of notice is usually stated in the contract. There may in fact be two periods of notice, such as fourteen days notice of intention to stop work until the payment has been made, or twenty-eight days notice of intention to terminate the contract on the grounds of failure to pay. Similar provisions may also exist for failure by the engineer to issue a certificate of payment.

What rights has the contractor if he disagrees with the amount stated by the engineer in a certificate for payment?

The contractor's right is to apply to the engineer for a correction or modification or adjustment to be made and included in the next certificate. If the engineer declines to do this, the contractor may either accept this decision or else treat it as a decision of the engineer with which he disagrees, and refer the issue to arbitration.

What the contractor should *not* do is to treat a disagreement over an amount certified as if it were a breach of contract by the employer: no breach of contract is involved, because there may have been a genuine error or a genuine ground for disagreement or difference of opinion. If so, then there should be, in most engineering contracts, procedures for dealing with such errors or differences, and the procedures must be followed. In the case of *Lubenham Fidelities and Investment Co. Ltd v. South Pembrokeshire District Council and Wigley Fox* (1986), where liquidated damages were wrongly deducted from an interim certificate, the contractor terminated work under the contract. It was held that it was not the employer, but the contractor who was in breach. The employer had made payment of the amount certified, and the error of the engineer could be subsequently corrected. The contractor had no right to terminate on these grounds.

What payment is due to a seller or contractor if a purchaser or employer terminates the contract or cancels any part of the order?

Several possibilities exist here and they should be anticipated and analysed at the time of the making of the contract. They may be looked at under the following headings:

1 termination due to the contractor's default;
2 termination due to circumstances beyond the control of either party;
3 termination or cancellation for convenience;
4 variations or changes;
5 wrongful termination or cancellation.

So far as the first of these categories is concerned (contractor's default), in many cases no payment will be due at all. Payment will not be due if it is an 'entire' contract and has not been completed at the time of termination. Even if it is not an entire contract, and there are stage payments already owing to the contractor, it may be that deductions to be made on the grounds of losses and expenses caused by the contractor's default and by the fact of termination will outweigh the payments due. In some cases, however, payment will be due to the contractor, notwithstanding the contractor's default, because the contract is a divisible one, and because partial payment has been earned and the damages due to the employer are not great enough to offset entirely the sums due to the contractor.

In engineering contracts, where termination occurs due to circumstances beyond the control of the parties, the basic principles are that the parties must look to the conditions of contract to see if payment has been expressly provided for. It may, for example, be provided that the employer must pay the contract value of all works executed at the date of termination, taking into account stage payments already made. But there is no common law rule about this, and if there is not an express clause in the contract to cover the situation, then we must ask whether or not the contract is a divisible one. If it is, then the employer must pay for those parts of the contract that have already been performed. If it is an 'entire' contract, then the contractor cannot claim the contract price as such, but he may claim for expenses reasonably incurred up to the date of the event which has rendered performance of the contract impossible, and he may claim a reasonable sum for any benefits conferred on the employer.

These statutory principles, as to expenses and remuneration for benefits conferred, only apply if the performance of the contract can be said to be 'frustrated'. A frustrated contract is something rather more serious than one affected by circumstances beyond the control of the parties. Circumstances beyond control, sometimes referred to as *force majeure*, may be only temporary in effect. Frustration is when such circumstances make performance of the contract impossible, or so radically different that for commercial purposes it can be called impossible. If there is *force majeure*, but not so as to amount to frustration, then the termination of the contract is optional, and not inevitable, and so the law does not make any provisions for payment. Such arrangements would have to be made between the two parties.

Termination or cancellation for *convenience* is a different type of termination or cancellation from that caused by *force majeure* or by frustration. The difference is that the employer reserving such a right may exercise it regardless of the circumstances, and does not have to give any reasons for the termination or cancellation. Such rights do not normally exist in engineering contracts, and are likely in the majority of cases only to exist because of specific terms in the employer's conditions of contract. Such clauses, which in effect give one-sided options to the employer, confer potentially valuable benefits upon employers, who may use them to terminate contracts or to cancel, wholly or partially, orders for goods and services: for example, they may decide that a project is not commercially viable, or that goods can be obtained cheaper from another source. Contractors, faced with such conditions of contract, will either reject them at the negotiating stage, or will at the very least look carefully at the balance of the relevant clauses to see what terms of payment apply in the event of such rights being exercised. Payment terms in such cases vary a great deal: some provide for payment only for goods actually delivered; others provide for reimbursement of expenses for work in progress as well; while yet another possibility is that there may also be provision for payment of compensation for any profit lost by the contractor due to the cancellation of work in progress. It is up to each party to negotiate the most favourable terms possible.

If termination or cancellation of any part of an order is made by the employer on grounds other than those described, and if the cancellation does not fall within the definition of a variation order, then the termination or cancellation will be wrongful, that is to say, it will amount to a breach of contract on the part of the employer. The employer who acts in such a way will be liable to make payment to the contractor of all sums which are already due by way of stage payments under the contract, as well as damages for breach of contract. Alternatively, even if the contract is an entire contract, the contractor can sue for a *quantum meruit*, that is, a fair and reasonable sum for the work performed. The law permits this because the contractor is entitled to a proper opportunity to earn money due under a contract.

Damages for breach of contract would be calculated on the basis of wasted expenditure and loss of profit by the contractor. Where goods are involved, the contractor may be able to mitigate loss by finding other markets for the goods, and if this is the case, the money raised by sale to other parties must be brought into account.

What deductions by the employer are allowed from payments?

As has already been seen, deduction by way of a retention is only permitted when the contract expressly provides for this. Other common provisions allowing for deduction to be made from payment by the employer are 'liquidated damages' clauses. These are clauses which provide for an agreed scale of damages in respect of certain types of breach on the part of the contractor. It is quite usual to find in engineering contracts clauses allowing for liquidated damages for delay by the contractor in completing the work. Such clauses are also found in sale of goods contracts and relate to delay on the part of the contractor in delivering the goods. Further details of this will be given in the next chapter, on delivery. Liquidated damages are sometimes provided, in engineering contracts, to cover the possibility that an item, or works, may not give the level of performance required. To give a simple example, if production equipment is sold, delivered and put up according to a specification which requires a given output of a particular end-product, it is possible for the parties, if they so wish, to agree that, within certain limits, any shortfall in the performance of the equipment could be compensated for by deduction of liquidated damages. If this is done, it brings an element of certainty and freedom from litigation into the contract, but it is advisable to pay close attention, when agreeing the contract terms, to the precise calculation and prediction of the likely losses to the employer resulting from the degree of shortfall in the performance of the equipment, extrapolated over an agreed period of time.

What is the position if there are no express provisions for deduction in respect of delay or defects?

In the absence of liquidated damages, deductions can still be made by way of set-off or counterclaim, but there will be much less certainty as to what the amounts to be deducted are, unless these are agreed by the parties, and consequently the chances of arbitration or litigation will increase. A typical example of set-off of this kind arises under the Sale of Goods Act 1979, section 53, under which the buyer may, if he wishes, elect to keep defective goods instead of rejecting them, and pay a lesser price on account of the defects. Although this is a perfectly legitimate option open to the buyer, the practical drawback is in deciding what is a fair price for the defective goods, and what is the amount to be deducted. A right of set-off only permits deduction of the correct amount, and one of the advantages of liquidated damages is greater certainty as to the amount to be set-off. Set-off arises from statutes and from principles of equity. If sums claimed on each side are *liquidated* (that is, capable of being ascertained with certainty at the time of the set-off), then they can even arise from different contracts: for example, in the case of *Axel Johnson Petroleum AB v. M. G. Mineral Group AG* (1992). If, on the other hand, one of the debts is unliquidated, then set-off is only allowed as long as the two debts, such as the price and a claim in respect of delays or defects, arise out of the same contract.

Can the rights of set-off and counterclaim be excluded by the terms of contract?

Parties are free to make arrangements between themselves as to payment, and this freedom includes the right to make their own accounting arrangements, and arrangements for dealing with cross claims. So, in theory, it is possible that terms of contract may exclude rights of set-off and counterclaim. In financial contracts this is fairly common, since an account holder may also be a debtor under a different account with the same lender or bank. A clause excluding set-off would make commercial and accounting sense, so as not to confuse two sets of accounts. In the case of *Hong Kong and Shanghai Banking Corp. v. Kloeckner & Co. AG* (1989), an agreement to pay the sums advanced to the bank, without deduction or set-off, was upheld as valid by the High Court. The borrower was seeking to set-off sums owing to it under a standby letter of credit, but was not allowed to do this.

On the other hand, in a United Kingdom contract, a clause excluding set-off or similar remedies will be looked at by the courts to see if it is reasonable and has any commercial justification. In the case of *Stewart Gill Ltd v. Horatio Meyer Ltd* (1992), there was a contract for the supply and installation of an overhead conveyor for the price of £266 400, to be paid in stages, with the final stage of 10 per cent of the price being payable in two instalments of 5 per cent on completion, and 5 per cent thirty days thereafter. On completion, the buyer withheld the final 10 per cent of the price on the grounds of breaches of contract by the contractors. The contractors sued for the money, and applied for summary judgment on the grounds that set-off and counterclaim were not permitted under the conditions of contract, which were the contractor's standard terms of business. One of these terms stated that:

> the Customer shall not be entitled to withhold payment of any amount due to the Company under the Contract by reason of any payment credit set-off counterclaim allegation of incorrect or

defective goods or for any other reason whatsoever which the Customer may allege excuses him from performing his obligations hereunder.

The Court of Appeal held that the particular clause in the contractor's conditions of contract was an unreasonable clause. The courts have power to disallow unreasonable conditions of contract under the Unfair Contract Terms Act 1977, and in this case it was Section 13 of that Act that applied, under which a court can declare a term of a contract to be unreasonable if it makes a liability or its enforcement subject to restrictive or onerous conditions or excludes or restricts any right or remedy in respect of the liability, or excludes or restricts rules of procedure. This means that an exclusion of set-off or similar remedies can be held to be unreasonable where there is no commercial justification for it, or where the condition of contract in question is too wide and too restrictive.

The Government, in its consultation paper issued in 1995, titled *Fair Construction Contracts*, has noted that the subject of set-off is of considerable importance in construction contracts, and that over-use of set-off is potentially damaging, as well as exclusion of it. The aim of the consultation is to try to establish a consensus about this, among other issues. A further comment on this consultation paper will be made later.

4 Terms about risk and delivery

Risk

The meaning of risk

All commercial contracts involve an element of risk, and one of the important functions in preparing for engineering contracts is the assessment of the risks, and the use of various legal and commercial techniques of managing the risks. 'Risk management' is the art of applying management skills to minimize the risks in a commercial project.

There are, of course, technical and financial risks (as well as, sometimes, logistic, geographical and political risks). In this chapter we are concerned with the need to arrange for the safe delivery of goods and materials within the agreed time. The expression 'risk', in this particular context, has always had a special meaning in law. 'Risk', in its true legal sense, as found for example in the Sale of Goods Act 1979, is the liability to bear a loss, should it occur, due to the loss of, damage to, deterioration of, or perishing of goods. Since engineering contracts, other than those which are purely about research and development, usually involve goods and materials, it is normal to find provisions in such contracts about which party is to bear the risk, at any given time, in this particular sense. There are other meanings of the word 'risk', and it is possible that in certain forms of engineering contract there may be different circumstances defined as either being the risk of the purchaser, or else as the risk of the contractor. These may include risks relating to use of a site, or about design, or use of certain works. Such clauses in contracts must be read with great care, as they give new definitions to the word 'risk'. However, in this chapter, the word 'risk' will be used mainly in its primary legal sense, that is, in the sense of risk to goods.

Risk may arise *before* delivery, *during* delivery, or *after* delivery. If goods are affected by a risk, the result may be financial loss, not only because of the cost of the goods themselves, but also due to loss of time, recovery costs, and transport costs, among other things. The aim of a good contract is to make the parties aware of what the risks are, so as to control them as far as possible, by good systems of packaging, transport, storage, security, and by any other appropriate measures. The contract should also make it clear where a loss actually lies from a legal point of view. This not only avoids unnecessary litigation, but should also lead to quicker acceptance of liability by insurers, with quicker settlement of claims.

Risk and insurance

It is always important to know that goods and materials are insured. Insurance alone, however, is not enough in an engineering contract. Both parties may have some form of insurance, but it should be clear which party's insurer is liable to pay if a loss should arise. From this point of view, a good engineering contract should have not only provisions about the duty to insure and the type and amount of insurance required, but also provisions about which party bears the risk in goods and materials at any particular time. At first sight it might seem academic which party is insured in respect of goods, as long as one or other of the parties is insured. In fact, this is not the case for a number of reasons. To begin with, there are hidden costs in making a claim on one's insurance: there is the cost of administration, a possible delay in receiving the cash in settlement, and the possible effect on the terms of one's insurance in the future, as a result of having made a claim. Next, there is the possibility of under-insurance, and the application of the law of 'average' (that is, the law that settlement is reduced in proportion to the amount by which the goods are under-insured). If you are the party at risk, and if you are under-insured, then you will bear the loss caused by the shortfall in the amount paid in settlement of a claim. If the other party is the person at risk, then the value of the goods for insurance purposes is the problem of that party. Similar considerations arise where there is an *excess* on the insurance policy, or where the loss occurs due to a risk which is excepted under the term of the policy. Finally, it is important to know who bears the risk because of the laws of 'subrogation'. Subrogation means that an insurance company can make use of the rights of the insured party, if a claim arises and is settled with the insured. Suppose that Company X has the risk in a cargo of goods, and the goods are damaged due to a breach of contract by Company Y, which is handling the goods for Company X. If Company X is the party at risk in respect of the goods, the insurer of Company X will not only be the insurer which has to pay for the loss, but will also be the insurer which then has the right under the law of subrogation to sue Company Y for damages for breach of contract.

Where does the risk in goods lie?

The incidence of risk depends upon what the contract between the two parties actually provides. If nothing is stated, the statutory rules, as well as rules of common law, apply. For example, in a contract for the sale of goods, the Sale of Goods Act 1979, section 20, states:

(1) Unless otherwise agreed, the goods remain at the seller's risk until the property in them is transferred to the buyer, but when the property in them is transferred to the buyer the goods are at the buyer's risk whether delivery has been made or not.
(2) But where delivery has been delayed through the fault of either buyer or seller the goods are at the risk of the party at fault as regards any loss which might not have occurred but for such fault.
(3) Nothing in this section affects the duties or liabilities of either seller or buyer as bailee or custodier of the goods of the other party.

The position in an engineering contract is less clearly marked out, partly because the law on engineering contracts is not codified in the way that the law of sale of goods is, and partly because the position as regards ownership of goods at any particular time is less easy to pin-point than in a sale of goods contract. For this reason, engineering contracts usually contain an express clause, giving the parties' chosen position, as regards the risk in the goods and materials with which the contract is concerned.

Terms about delivery

From a purely functional point of view, terms about delivery are primarily to determine how and where goods are to be delivered, and at what time. They may also include further details such as packaging, marking, loading and unloading, and inspection on arrival. From a legal point of view, such terms place responsibility upon one, or sometimes both, of the parties. Terms about delivery also have close links with terms about risk, title to goods, insurance, payment, and other matters. So close are these links, that a body of assumptions has grown up around some terms of delivery, so that to select a form of words, such as 'Free on Board', or 'Delivered at Frontier', is to make a choice about these matters, unless the contract already contains specific provisions to the contrary.

There are three basic ways in which terms about delivery can be set out in a contract.

(a) The parties may make their own arrangements, using their own form of words, setting out their own positions as agreed.

58 *Terms about risk and delivery*

(b) Alternatively, the parties may make use of expressions (and corresponding abbrieviations) recognized by law. If, for example, they state that goods are to be delivered 'ex works', or 'CIF', each of these expressions has an understood meaning in English law. In English law, implied terms about the place and manner of delivery, and about risk and title, have developed over a period of a century or more, so that although the parties are still free to some extent to agree their own specific terms, the use of a recognized expression will make it unnecessary to do so unless the parties wish to create a variant on the standard position.

(c) Alternatively, the parties may wish to make use of and to refer to INCOTERMS in their contract. INCOTERMS are a set of conditions of carriage which are produced by the International Chamber of Commerce, which are revised from time to time, and if parties to a contract intend to refer to them they should be sure that the version referred to is clearly stated. The current version is INCOTERMS 1990.

INCOTERMS make use of the traditional trade terms and practices, but have a number of important differences. If, for example, the expression FOB (Free on Board) is referred to in a contract to which English law applies, the meaning of the expression will be deduced from the surrounding terms of the contract, and by reference to precedents on the law relating to carriage of goods. If, on the other hand, the use of the expression FOB is coupled with a statement that INCOTERMS 1990 will apply to the contract, then the meaning of FOB, and the respective duties of the parties will be set out in the INCOTERMS themselves. INCOTERMS have similarities to the rules that would apply under English law, but are more explicit, and are set out as a series of ten duties laid upon each party, in a stated sequence. For each INCOTERM the duties are different. They aim to reflect practices common to most of the well-known systems of law, and are therefore highly suitable for use in the appropriate international transactions.

Table 4.1 shows the conventional English law terms for use in contracts involving the carriage of goods. It also shows the corresponding INCOTERMS, where they exist.

What do the different terms about delivery mean?

To set these out in full would be beyond the scope of a work of this kind, so only an outline of a few of the different terms will be given. The observations that follow are about the conventional terms of delivery as interpreted by the Common Law, and are not intended to set out or to describe INCOTERMS. Those wishing to incorporate INCOTERMS into contracts should consult the booklet of INCOTERMS produced by the International Chamber of Commerce.

Table 4.1

Conventional terms	INCOTERMS 1990
Ex Works	EXW – Ex Works
Free on Rail – FOR	
Free on Truck – FOT	FCA – Free Carrier (named place)
Free on Board (Airport) – FOB	
Free alongside Ship – FAS	FAS – Free Alongside Ship
Free on Board – FOB	FOB – Free On Board
Cost and Freight – C&F	CFR – Cost and Freight
Cost, Insurance and Freight – CIF	CIF – Cost, Insurance and Freight
	CPT – Carriage Paid To
	CIP – Carriage and Insurance Paid To
	DAF – Delivered At Frontier
Ex Ship	DES – Delivered Ex Ship
	DEQ – Delivered Ex Quay
	DDU – Delivered Duty Unpaid
	DDP – Delivered Duty Paid

Ex works

Delivery *ex works* means that the seller of the goods has to make the goods available to the buyer or the buyer's carrier at the seller's works or factory or other premises. It is then up to the buyer to transport the goods. Responsibility for loading the goods onto the vehicle of the buyer is that of the buyer, unless otherwise agreed. Risk and the property in the goods pass to the buyer as soon as the goods have been placed at the disposal of the buyer, unless otherwise agreed.

FOR/FOT

These initials mean *Free on Rail* and *Free on Truck* respectively. 'Truck' in this context means a railway wagon, so the expressions are only to be used where rail is to be the means of carriage. The INCOTERM which may be used instead of these expressions, and which will cover any mode of transport, whether rail, road or air, is FCA (Free Carrier). Under the FCA term, the seller does not have to load the goods onto the mode of transport, but only has to deliver into the custody of the carrier or other named person at a named point, such as a transport terminal.

FAS

This is *Free Alongside Ship*. Under a contract made on these terms, a seller must place the goods alongside the designated ship, on the quay or at the loading place named by the buyer. At this point, unless otherwise stated, the risk and property, and all further costs involved in export and carriage are with the buyer.

FOB

This well-known set of initials stands for *Free on Board*. The seller's duty is to deliver the goods to a port of shipment named by the buyer, and to have the goods placed on board a ship nominated by the buyer, for the account of the buyer, and to procure a bill of lading in terms usual in the trade. This was described in the case of *Pyrene Co. Ltd* v. *Scindia Steam Navigation Co. Ltd* (1954) as the 'classic' type of FOB contract. However, in modern times the FOB contract has become a flexible instrument, and any number of different variations on the classic type may be made. In some cases the seller may be asked to make the necessary shipping arrangements, instead of the buyer, and the bill of lading may be in the seller's name. In other cases the buyer's forwarding agent will book the space on the ship, and the seller will place the goods on board, and will obtain a mate's receipt, and will hand this to the forwarding agent, so as to enable the forwarding agent to obtain a bill of lading. Risk in FOB contracts is generally on the buyer when the goods have been lifted over the rail of the ship. The seller must give notice to the buyer to enable the buyer to take out insurance of the goods during sea transit. FOB is unsuitable for contracts where the ship has no 'rail', and 'roll-on, roll-off' methods are used. The INCOTERM, FCA, is more suitable for such contracts, and is also intended to replace FOB Airport terms.

CIF

These initials stand for *Cost, Insurance and Freight*. With a CIF contract, the seller must make out an invoice for the goods to be sold, ship goods of the contract description at the port of shipment, procure a contract of carriage by sea, and pay the costs and freight necessary to bring the goods to their port of destination. He must also arrange for insurance on terms current in the trade, such insurance to be available for the benefit of the buyer. The seller must tender to the buyer the bill of lading, invoice and insurance policy. If the provision of insurance is to be by the buyer instead of by the seller, then the CIF form of contract will be inappropriate, and instead a modification of the terms, known as C & F (Cost and Freight), or CFR, (if INCOTERMS are used) will be employed by the parties.

With CIF, risk in the goods passes on shipment. The property passes according to the terms of each particular contract, but if nothing is stated to the contrary, then property passes when the documents are handed over to the buyer. If roll-on, roll-off, container transport is used, it is probably better, instead of the CIF form, to use the CIP (carriage and insurance paid to) terms provided in INCOTERMS.

Ex ship

Here, the seller must arrange for delivery to the buyer at the port of delivery. The seller must therefore arrange and pay for carriage by sea and insurance. Risk and property remain with the seller until the goods are delivered to the buyer. Unlike a CIF contract, payment is not against documents, and the buyer is not bound to pay until the buyer has the goods (subject to any other terms stated in the particular contract). Removal of the goods from on board the ship, and clearance of the goods, is for the buyer to carry out.

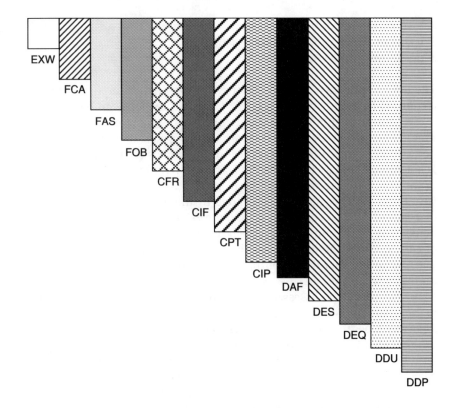

Note: This is a comparison of the extent of the duties undertaken by the sellar in the different terms of delivery in sales of goods.

Figure 4.1 *Methods of delivery*

Ex quay

This is similar to ex ship, except that the seller must make the goods available to the buyer on the quay or wharf at the port of destination. If the INCOTERMS 1990 version is used, then the seller must arrange for the goods to be cleared for importation, and all duties, taxes and other costs up to the point of delivery must be borne by the seller.

The meaning of times for delivery or performance

Whenever times are given in a commercial contract for the carrying out of certain obligations, a legal question immediately arises: the question is what legal significance is attached to the time or times, or date or dates, as the case may be. Such times are capable of being either:

1 essential terms of the contract, or
2 terms of the contract, but not essential, or
3 not terms of the contract at all.

The normal phrase used to make time an essential or important term of a contract is 'time is of the essence'. Other expressions can be used, and if the context is appropriate, they will have the same effect. By making time 'of the essence', the parties are agreeing that the time or date for performing a particular duty under the contract is so important that if it is not strictly kept to, the party in breach will have broken a material term of the contract. The result will be to give to the other party the right to treat the contract as having been ended by the party in breach. In the case of *Lombard North Central plc v. Butterworth* (1987), there was a leasing agreement for a computer, in which it was stated that time for payment of each quarterly rental was 'of the essence'. The Court of Appeal held that this meant that if the contract was terminated for arrears of rentals, the *customer* would be the party who had terminated the agreement. Damages would be assessed on this basis, and would be significantly higher than if it had been the leasing company which had terminated the agreement, because *all* the loss to the company could be said to have been brought about by the act of the customer. If time for payment had been a term of the contract, but not 'of the essence', then the result would have been different. The company would still have been able to enforce the term, but only after giving notice to the customer; if the customer had then failed to pay, the company would have been able to terminate the agreement and to obtain damages, but the scale of damages would have been lower, because the decision to terminate would have been that of the company, and so the company would not have been able to claim that the customer caused the same amount of loss to it as if the customer had terminated the agreement.

In the case of *Kolfer Plant Hire Ltd* v. *Tilbury Plant Ltd* (1977), it was inferred by the High Court that the use of the phrase 'at the latest', coupled with the fact that the customer asked for a diesel generator by a specific date, meant that time for delivery was of the essence. This entitled the customer to reject the generator when it was not delivered on time, even though it was sent only one day later than required. If time is of the essence, only the smallest amount of delay will amount to a breach of the term, and no notice by the other party is needed before treating the contract as ended. In this case it was held that the customer not only had the right to reject the generator, but was also entitled to damages for the breach.

It is most likely that it will be inferred that time is of the essence when a definite and fixed time is given for performance of an obligation. However, even a less definite period of time can be held to be of the essence, if the context so requires. In the case of *McDougall* v. *Aeromarine of Emsworth Ltd* (1958), the contract was for the building of a boat. A clause in the contract stated that the builders would use their best endeavours to complete by 1 May 1957, but that the delivery date could not be guaranteed. (Such expressions should be avoided if possible, as they are by no means easy to interpret.) In the event, the boat was not ready by September 1957, and the buyer decided to terminate the agreement and to claim a refund of payments already made. The court held that the buyer was entitled to do this without further notice, since time was of the essence. At first sight this judgment seems to contradict the terms of the contract, but the reasoning is that 'best endeavours' has a meaning, and although the qualifying words made the date of 1 May approximate only, there was a duty to make all reasonable effort to deliver within a reasonable time from that date, and that to that extent time was of the essence.

In building and engineering contracts, particularly for special or sub-contracted works, such as installation of equipment or provision of electrical or electronic works, it is not uncommon to find that time for delivery or completion is made of the essence. It is, however, less common in building, construction, or civil engineering, since such a term creates too great a risk for the contractor. In a long-term contract of this kind, it would be most unwise for the contractor to agree a term under which, for the slightest delay in completion, the contract could be terminated without notice. The employer or purchaser would not necessarily benefit from such a term, either, since one of the problems about time being of the essence in such a contract is that the purchaser must wait until the due date for completion before being able to say that there is a breach of contract. In many cases the purchaser is well aware that progress is being delayed, long before the due date for completion of the works has arrived, and would like to be able to take action. Making time of the essence does not necessarily have the required effect.

In building and engineering contracts, rather than making time of the essence, the purposes of both parties are better served if the contract provides for a detailed programme which the contractor is to meet. If this programme is linked to terms of payment, then the cash-flow incentive of the contractor to make progress is clear enough. Further, the contract can always provide for 'liquidated damages' in the event of a delay by the contractor. The point about liquidated damages is that the contractor who is warned that they will be applied if he is late is more likely to take remedial measures to reduce the delay, rather than suffer the loss of money by way of deduction of liquidated damages. Finally, the contract can always achieve the last resort effect of time being of the essence by means of a clause entitling the purchaser to terminate the contract if the contractor fails to make due and diligent progress. The benefit of such a clause to the purchaser is that it can be applied *before* the date for completion under the contract. The benefit to the contractor is that notice would be needed, and the contractor is at less risk of having the contract terminated for a comparatively minor delay.

Liquidated damages

As mentioned in the preceding paragraph, liquidated damages are an option available to parties negotiating the terms of an engineering contract. the aim of liquidated damages is to provide for a negotiated and fair method of allowing for the possibility of a delay in completion of work, which provides adequate compensation for the purchaser, while ensuring that the contractor is not too heavily penalized for the delay. The word 'liquidated' means that the measure or scale of such damages has been set down in the contract in agreed and mathematically quantifiable terms. This usually means either an agreed daily or weekly rate, for example, 'LIQUIDATED DAMAGES: £1000 per week or any part of a week' or alternatively, the parties may agree to set down the liquidated damages as a percentage of the contract price or value, for example, 'LIQUIDATED DAMAGES: 0.5 per cent of the contract value per week, or any part thereof'.

There are many negotiable options for the parties in drafting a provision as to liquidated damages. Apart from the obvious difference between a monetary sum and a percentage of the price, there is also the question of whether the sum is to be payable on a daily or weekly basis. Then there is the question of whether 'week' means a completed week, or any part of a week. There is also the possibility that the parties may agree upon a period of grace before the liquidated damages apply. Contractors also need to be aware that there are no limits to the length of time for which liquidated damages can be applied unless these have been set down in the contract. So if the scale is £1000 per day and the contractor is 100 days late, the sum of £100,000 may be deducted from the amount due to the contractor, unless the contract provides for a lower limit than this, to the amount of the possible liquidated damages. Parties to the contract should also note that liquidated damages may only be applied in the precise manner stated in the contract. Therefore, if the contract says that they may be deducted from the final payment, this means that they cannot be deducted from interim payments. It also means that they can only be *deducted* from sums owed to the contractor, and not recovered from sums already paid to the contractor. On the other hand, the words, 'the Contractor shall pay or allow to the purchaser by way of liquidated damages the sum' would have the effect of permitting the purchaser not only to deduct liquidated damages from sums owing to the contractor, but also to recover the liquidated damages by suing for them, notwithstanding that the contractor may already have been paid.

Summary of key points to negotiate with liquidated damages

- Rate
- Style (monetary sum or percentage)
- Time basis (daily, weekly, monthly, etc.)
- Period of grace (if any)
- Limit of liability
- Method of application
- Other relevant points (for example, relationship to the remainder of the contract)

On the last of the above points, it should be mentioned that there is nothing to prevent a purchaser from insisting that time be of the essence *as well as* including liquidated damages in the contract. A contractor should, however, resist this wherever possible, since it cuts across the whole commercial point of having a provision for liquidated damages, which is to agree a fair division of risk of delay. Another point which may arise is as to what the position is, if the contract contains a limit to the number of days or weeks for which liquidated damages can be applied. For example, the contract may provide that the liquidated damages may be paid to, or deducted by, the purchaser at the rate of £1000 per week of delay up to a maximum of ten weeks. Once the contract is in delay by ten or more weeks, the common law position will be that the purchaser will be entitled to give the contractor notice to complete within a reasonable time. If this time is set, and is not complied with, the purchaser will be entitled to terminate the contract. The question of whether the purchaser will then be allowed to claim *further* damages, over and above the liquidated damages, is a complicated one, and may depend upon the exact wording of the contract. There are several engineering contracts which provide for this eventuality.

The rule against penalties

When two parties agree upon liquidated damages, at the time that a contract is made, they are, presumably, satisfied that they have struck a reasonable bargain. Parties to commercial contracts do not enter into them with the intention of challenging the terms at a later date. However, with liquidated damages, such challenges are not unknown, and where this occurs, it is because in this area of the law there exists a particular rule, known as the rule against penalties. This rule does not exist in every legal system, and should only be considered in systems which have adopted a form of the common law. This rule is that where a contract provides for damages to be paid by a party in breach of contract, the amount or scale of damages provided for must be such that it is intended reasonably to compensate the innocent party, and not to punish the party in breach. If the amount or scale is excessive or punitive, then it is a 'penalty'. A penalty is null and void and of no effect: the contract will simply be read as if the clause is not there.

A great deal of unnecessary confusion still exists as to what is the precise test to distinguish between genuine liquidated damages, which are valid and enforceable, and a penalty, which is not. The reason for this difficulty is that often parties to a commercial agreement will find that when the time comes to apply the agreed liquidated damages, the actual loss to the purchaser is rather less than the liquidated damages applicable to the delay in question. In many cases the purchaser may have been delayed by other matters, as well as being delayed in relation to the particular contract. The purchaser may have revised his internal schedules and requirements, without changing the programme for the particular contract. In such cases, many contractors will, naturally, think it unjust that the purchaser should claim the liquidated damages in full, even though the loss has not materialized to that extent.

However, the perceived justice or injustice to the contractor, at the time at which the liquidated damages are applied, that is, at the time of completion of work, is *not the correct test of the legality of the liquidated damages*. The true test is what the parties are presumed to have had in mind *at the time of the making of the contract*. At that time, we must ask whether or not the scale of liquidated damages was genuinely negotiated and agreed as an attempt to pre-estimate, in good faith, the likely loss to the purchaser. If, taking into account the wording of the contract, the scale of the damages set down in the contract, and the basis on which the pre-estimate of the likely loss to be suffered by the purchaser was calculated, the measure or scale appears to be reasonable, then the sum will be treated as liquidated damages. If not, then it will be treated as a penalty. Simply calling a clause or sum 'liquidated damages', will not save it from being treated as a penalty, if that is the true position after applying the test as to what is genuine liquidated damages. It is, of course, unwise to use the word 'penalty', because although the word as such is not conclusive, it will create a presumption that it is meant to penalize rather than to set liquidated damages. No better explanation of the rule against penalties has been given than that of Lord Diplock in the case of *Photo Production Ltd* v. *Securicor* (1980):

> Parties are free to agree to whatever exclusion or modification of all types of obligations as they please within the limits that the agreement must retain the legal characteristics of a contract and must not offend against the equitable rule against penalties; that is to say, it must not impose upon the breaker of a primary obligation a general secondary obligation to pay to the other party a sum of money which is manifestly intended to be in excess of the amount which would fully compensate the other party for the loss sustained by him in consequence of the breach of the primary obligation.

The rule against penalties applies to virtually every type of contract, and not only to engineering contracts. Sales, hire, loans, contracts of employment, agency, distribution, and joint venture contracts have all at some time or other produced examples of the application of the rule. It follows that great care is needed in drafting any clauses in commercial contracts which fix a sum or scale of damages for breach by either of the parties.

Some legal questions answered

Can liquidated damages be applied to separate parts or portions or phases of work to be done under an engineering contract?

The answer to this question is quite simply, yes, provided that the contract makes it clear how the damages are to apply, and provided that the sums or scale as apportioned still pass the 'genuine pre-estimate' test. It is perfectly possible for the relevant clause to refer to a percentage of the value of the parts of the works which cannot be put to use in consequence of the delay. What must not be done, however, is to attempt any kind of apportionment which is not specifically provided for by the contract. In JCT (Joint Contracts Tribunal) Building Contracts, which are sometimes

68   Terms about risk and delivery

used for engineering works which are to be done as part of a building project, or to be incorporated into a building, there are special forms for 'sectional completion', which the parties may use where appropriate. In the case of *Bramall and Ogden Ltd* v. *Sheffield City Council* (1983) the purchaser took over buildings built under the JCT contract in sections. The contract was for 123 houses, and a sectional completion form had not been used. It was held that the purchaser was not entitled to deduct *any* liquidated damages under this contract, since, firstly, the overall figure of liquidated damages would be too high for work taken over in sections, and an apportioned figure was not possible, even though the purchaser would have been happy to have done such an apportionment, because the contract made no provision for it.

What is the position if the contract provides for liquidated damages which have the effect of undercompensating the purchaser?

The rule against penalties, as such, will not apply in such a case. Penalties are always an excessive sum, and not a sum which is too lenient. In *Widnes Foundry Ltd* v. *Cellulose Acetate Silk Co. Ltd* (1933), a contract was for the delivery and construction of an acetone recovery plant. The liquidated damages were very low, and after a delay of thirty weeks, the buyer attempted to sue for approximately ten times the amount of the liquidated damages that would have applied. It was held that this was not possible, as the circumstances were fully covered by the liquidated damages clause.

What is the liability for delay in delivery or completion of work if there is no liquidated damages clause?

If time is of the essence, then the liability is that the contract may be treated as terminated, and damages can be sought by the purchaser. Alternatively, the purchaser may make use of one of these remedies without the other. The damages would be those actually suffered by the purchaser, and would reflect heads such as: loss of use; loss of production; additional costs, such as those of having to take the work to a different contractor; increased costs, if the item to be delivered has to be obtained from a different source at a higher price, etc. In the case of *Koufos* v. *Czarnikow* (1967), which although not an engineering case, is a leading authority on the law of Sale of Goods, it was held that a buyer who buys intending to resell the goods, where the circumstances were known to the seller, or ought to have been appreciated by the seller, can claim damages for loss of profit if the goods intended for resale are delivered to him after the due date for delivery, and if the resale market falls subsequently.

What is the position if a liquidated damages clause is thought to be a penalty and is challenged by the contractor?

This is the most likely scenario in which the rule against penalties might come into play (although there have been cases when it has been the customer who has invoked the rule, or cases when the damages have applied equally to both parties, as in a partnership or joint venture). The contractor who wishes to challenge liquidated damages at the time they are to be applied must think out his position very carefully, and do accurate calculations. The position is that if the liquidated damages clause is successfully challenged, the contract is read as if the clause is not there. This does *not* mean that no damages are payable. What it means is that damages are set at large, and can be claimed by the party not in breach on the basis set out in the previous paragraph: that is, as if there had never been a liquidated damages clause in the first place. If the actual loss suffered by the party claiming is lower than the liquidated damages would have been, then there is likely to be some advantage in making such a challenge. If, on the other hand, the liquidated damages served to limit the liability of the contractor (or the party exposed to the liquidated damages, if not a contractor), then there would be no purpose served by challenging them, since the challenge would not lessen the liability, even if the clause were to be struck out, and might even increase the liability in some cases.

Is a liquidated damages clause incompatible with time being 'of the essence'?

If the parties want to agree terms under which time is of the essence, and under which, as an additional option to the purchaser, there are also liquidated damages for delay, there is nothing to prevent them from doing so. The possibility exists, and purchasers have occasionally driven a bargain of this kind. In some engineering contracts, it is no more than academic, whether one form of liability, or the other, or both, are stated in the contract, since the likelihood of delay is negligible. In other engineering contracts, where the circumstances make delays a considerable risk, the existence of both forms of liability would be unattractive to the contractor. It should be noted, however, that if liquidated damages are included in an engineering contract, and *if nothing else is said which expressly makes time of the essence*, then there will be a presumption that time is *not* of the essence. The reason for this is that by providing for liquidated damages, the purchaser is by implication making an allowance, in commercial terms, for the fact that the delivery of goods or completion of work may be late.

Is there a legal distinction between a delay and a defect?

Not all delays are due to defects, and not all defects have a relationship to delays, but the two concepts can overlap. If goods are delivered on time, but at the time of delivery are checked and inspected and counted by the purchaser, it may be that damage or discrepancies are immediately apparent. In such a case, the purchaser may at once reject part or the whole of the goods, depending upon the circumstances. If the goods can be replaced by the seller before the final date allowed by the contract for delivery, then there is no delay. If, on the other hand, the defective goods require some time to be replaced, then there were not only defects, but also delays. There would therefore be further remedies available to the purchaser for the delay. Similarly, if the contract is for engineering work to be tested upon completion, it may be that the contract provides for the rejection by the purchaser of work which fails the tests on completion. In such circumstances, it can be said that the defects will almost certainly cause a delay, unless the tests have taken place well in advance of schedule, or unless the defects are relatively simple to put right. Serious defects which require time to put right will probably mean that there will be delay in passing the tests and in handing over the works to the purchaser. Such delays are the type of delays that clauses about completion on time, and liquidated damages are intended to deal with.

On the other hand, there are at least three types of defects (and possibly more), which are unrelated to the concept of delay.

(a) There are defects which are noticed on a pre-delivery visit, test or inspection; these need not necessarily cause any delay at all, as they are often an internal matter.

(b) Next, there are those defects which, while they may be noted at the time that the tests on completion are carried out, do not affect the taking over and use of the plant, works or equipment in question. Such defects may then be either the subject of a waiver or concession, or alternatively may be listed as defects to be put right after taking over, under the defects liability or warranty provisions of the contract. A well-drafted and well-managed engineering contract will provide for this possibility, since it can be of benefit to both parties, providing security for the purchaser as well as allowing the contractor the benefits of a completion or taking-over certificate. The taking-over, or equivalent, certificate should be drawn up accordingly: see the example given in Chapter 2.

(c) Finally there are defects which are not apparent at the time of completion or taking over, but which materialize later on or are revealed by the Performance Tests which are carried out after taking over. These do not affect completion on time, and if they arise are simply to be put right under the warranty or other defects liability provisions. Consequently no liquidated damages for delay arise, although under some engineering

Terms about risk and delivery 71

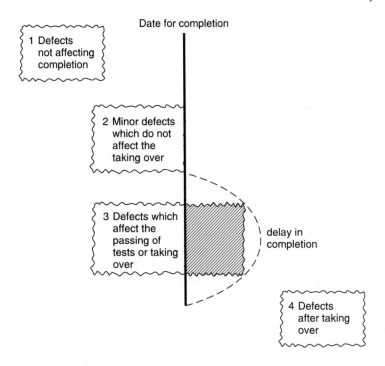

Figure 4.2 *Illustration of the effect of defects upon completion*

contracts a different form of liquidated damages may arise at this stage: these are liquidated damages, if any, for defects which cannot be put right, and which may cause a permanent shortfall in the output or performance of capability of the equipment. This form of liquidated damages, which has nothing to do with delay, and which is measured by entirely different criteria, was discussed in Chapter 2.

Is the liability of the contractor for delay in completion altered if it is caused by circumstances beyond the contractor's control?

This depends upon the circumstances, and upon the relevant conditions of the contract. In some cases, the circumstances may be beyond the contractor's strict control, but they will nevertheless be risks of a kind which the contractor should have foreseen and should have guarded against. Burglaries and vandalism of goods and materials which are under the contractor's control and at the contractor's risk will probably come into this category. A slightly different set of circumstances, but also part of the contractor's own risk in most cases will be defects in goods and materials delivered by a sub-contractor, which in turn cause the contractor to be late. This is a contractor's risk in the sense that the contractor should guard against this by the proper selection of the sub-contractor or supplier, as well as by legal and commercial means, and should, as a result of this, have arrangements under which the loss is passed down to the sub-contractor, after it has first been made good by the contractor to the purchaser. Delays by sub-contractors, and the insolvency of a sub-contractor are, in most cases, unless the contract makes specific exceptions, risks of the contractor. Shortage of materials or labour is usually a contractor's risk, unless the contract makes an exception for this.

However, most contracts provide for certain events which are beyond the contractor's control, and in respect of which the contractor will be entitled to an extension of time for completion. If the extension of time covers the full period of delay, then the contractor will have no liability for delay. If the extension of time covers less than the full period of the delay, then the amount of delay not covered will remain the liability of the contractor. The two main types of occurrence which give the contractor the right to an extension of time for completion are *variations* (which have already been discussed in an earlier chapter), and *acts or omissions of the purchaser*. Apart from these, the contract will normally allow the contractor to invoke any circumstances which fall within the definition of *force majeure* as a ground for extension of time. However, the right to invoke *force majeure* does not occur automatically: a fact that is not always appreciated by sellers and contractors. There is, for example, *no automatic right* to claim an extension of time simply because work has been affected by a strike or other form of industrial action. In order to be entitled to claim *force majeure*, the contractor must show that circumstances of the kind that have arisen are expressly, or by implication, mentioned in the contract, and that the delay is attributable to those circumstances. More will be said about *force majeure* and *force majeure* clauses in Chapter 5.

Formula

Formula for calculating the delay for which a contractor is liable after an extension of time.

$$\frac{\begin{array}{l}\text{actual period of time taken to complete work} \\ \text{minus: contract time or period for completion}\end{array}}{\begin{array}{l}= \text{delay in completion} \\ \text{minus: allowable period for circumstances beyond contractor's control}\end{array}}$$

= delay for which contractor is liable

5 Progress in engineering contracts

The programme

A few years ago it was possible to find engineering contracts, even contracts produced by major institutions, which had no contractually binding provisions in them about a programme. Such contracts relied purely upon a date for commencement and a date for completion. While this is possible with very minor contracts, the absence of a programme would raise problems in more significant engineering contracts because of the following requirements:

1 the need to monitor progress;
2 the need to provide and control site facilities;
3 the interaction of the programme with terms of payment;
4 the interaction of employer's and contractor's duties;
5 the interface between the work of one contractor and the work of other contractors.

Modern engineering contracts not only provide for a date for commencement and a date for completion, but they also provide for a form in which a programme must be submitted by the contractor, a date within which submission of the programme must be done, and a date within which the engineer is required to approve the programme. Different forms of institutional contracts do, however, differ considerably as regards the precise form of these clauses, and it should be noted that although most of them provide for the submission of a programme, they do not all provide for a specific time for submission or approval. Possibly, it may be argued that there are good reasons for being less than precise about such times; nevertheless, if parties wish to know exactly what is required of them in an engineering contract, there is a lot to be said for precise provisions about submission and approval of the programme, such as those to be found in the *ICE Conditions of Contract*, 6th Edition.

Where the contract does provide that a programme must be submitted and approved, then once it has been approved, it is a document with contractual effect, since most contracts go on to provide that the contractor must carry out the work in accordance with the programme, and that the contractor may not make any alteration to the approved programme without the consent of the engineer. Many forms of contract do, however, permit the engineer to order the contractor, on certain specified grounds, to revise the programme. A good engineering contract will distinguish between those revisions to the programme where the costs fall onto the purchaser, and those where the costs are to be borne by the contractor, are likely to be those revisions which have become necessary due to some act or omission or failure to make good progress on the part of the contractor.

It cannot be over-emphasized that without a programme containing sufficient detail, together with conditions of contract dealing with the other matters mentioned in the preceding paragraphs, an engineering contract is incomplete, particularly from a purchaser's point of view. Whereas, in a contract for the sale and purchase of goods, it may be sufficient to make time for delivery 'of the essence', this is *not* in itself sufficient in an engineering contract, since such a provision cannot be enforced until the date for completion has been passed. While the threat of enforcement may act as a form of disincentive to be late, the actual enforcement would come altogether too late in the day for the purchaser. Only a contractually binding programme which enables the purchaser to monitor progress on a regular basis and to point out discrepancies between the programme and the actual progress of the contractor, and to call for remedial measures, will suffice. Moreover, provisions such as making time 'of the essence', and providing for liquidated damages for delay, are very much *legal* and *adversarial* methods of approaching the problem of achieving completion of a project on time. While they undoubtedly have their place in commercial contracts, it is also fair to say that companies and other businesses will at times derive more benefit from more commercial and less adversarial methods of approach. Good programming provisions lend themselves to such an approach, without necessarily creating legal issues between the parties.

Force majeure

No contractual provisions, and no programmes, are proof against the forces of circumstances beyond the control of the parties. One of the most important features of English law is the need of the parties to commercial and engineering contracts to make provisions dealing with *force majeure*, or, if plain English is preferred, circumstances beyond the control of the parties.

Force majeure *and the law of frustration*

The common law did not originally recognize the idea of *force majeure*. What it did recognize, from the late-nineteenth century onwards, was the rather more limited principle of frustration of contracts. Frustration occurs where, due to a radical change in circumstances, beyond the control of either of the parties, performance of the contract becomes illegal or impossible, or so different from what was originally contemplated that the commercial venture is to all intents and purposes frustrated. No provision is needed in commercial contracts dealing with the possibility of frustration, since this is a rule of law which applies automatically. However, the circumstances which the courts recognize as causing frustration of contracts are very few indeed, and are probably limited to death of one of the parties, legislation making performance illegal, an outbreak of war making performance impossible, or physical destruction of the subject matter of a contract. Other possibilities exist but are irrelevant to most engineering contracts.

Circumstances which disrupt the performance of an engineering contract, but do *not* radically change it or make it illegal or impossible, are *not* frustration. However, the parties need, for obvious commercial reasons, to have a solution to the questions posed by such disruption. There may be delays due to bad weather, or due to floods, or due to the failure of public utilities, or due to industrial action, or due to civil commotion, to name a few examples. None of these are likely to be drastic enough to frustrate a contract under the common law. However, it is open to the parties to provide for such circumstances by means of a '*force majeure* clause' in the contract. If the parties do not have or refer to such a provision in their contract, the courts will not imply such a provision. But if the parties do have such a provision, the courts will certainly recognize it, and will give it effect according to its precise meaning. Many engineering contracts use the words *force majeure*, and then go on to give the words some form of definition. It is not necessary, however, to use those words, nor even to give definition to the circumstances. A clause entitling either or both of the parties to additional time on the grounds of 'circumstances beyond their control' will suffice in many cases, although it is obviously less precise in pinpointing the exact circumstances than a fully defined *force majeure* clause would be.

The occurrence of force majeure

There are two aspects to the use of *force majeure* clauses in engineering contracts. One is to form an understanding of the meaning of the particular clause and the circumstances that are covered by it. The second, when such circumstances are thought to have occurred, is to be well acquainted with and to observe carefully all the procedures that are required by the contract for the formulation of a valid claim for extended or additional time on the grounds of *force majeure*. This second aspect is particularly important from the point of view of the project manager or engineer.

Some engineering contracts require notification, in writing, by the contractor to the engineer, of the circumstances constituting *force majeure*, and of the claim for an extension of time, within a stated period of time (such as seven days). Other forms of contract use the more flexible words 'as soon as reasonably practicable'.

Force majeure *and breach of contract*

Force majeure, or circumstances beyond the control of the parties, does not involve a breach of contract by the parties, as long as the contract contains a suitable clause which makes this clear. The clause will usually contain a list or other form of definition of the relevant circumstances, and will go on to provide either for an extension of time for performance, or that neither party will be in breach of contract if delay is caused by any of the defined circumstances. Because no breach of contract is involved, the basic principle is that additional time is granted, if the required notice is given, but *no additional money is payable*. The contractor is not in breach, to the extent that performance of obligations is affected by *force majeure*, and the employer is not in breach, so that the effect is that each party must bear their own additional costs, if any, which arise as a result of the circumstances. Only if the contract were to make express provisions for additional payment, would such a claim be possible.

For this reason, *force majeure* (and frustration, where relevant), must be distinguished from *breach of contract*. Breach of contract is not *force majeure*, since it is not beyond the control of the parties. Nor is it the same thing as frustration: if a party to a contract were to make performance of a contract impossible, by missing a crucial deadline, this would not, in law, be frustration, even if as a result the contract had to be discontinued. It would be self-induced, and so would be classified as a breach, for which the party causing it would be liable to pay damages. Table 5.1 summarizes the distinctions between frustration, *force majeure* and breach of contract.

Claims for additional payment

As already discussed earlier, a claim for additional payment can be made by a contractor on the grounds of authorized variations to the specification, or work to be carried out under the contract. However,

Table 5.1 *Differences between* frustration, force majeure *and breach of contract*

Frustration	Force majeure	Breach of contract
Requires circumstances beyond the control of the parties.	Requires circumstances beyond the control of the parties.	The circumstances will have been brought about by one of the parties, in breach of an obligation.
Brings the contract to an end.	Usually allows an extension of time. If prolonged, can end the contract.	Whether or not it ends the contract depends upon the term, and upon how serious the breach is.
Financial consequences are governed by Act of Parliament. Only occurs if performance becomes illegal or impossible or radically changed.	Usually no financial adjustment.	Damages are payable for breach of contract.

engineering contracts often contain clauses about claims for additional payment to be made by the contractor, and the grounds upon which the contractor may make such a claim are often far wider than variations. They are dealt with in this chapter firstly because they often relate to matters affecting progress, and secondly, because they highlight the relevance of the distinction between breach of contract and *force majeure*.

Usually, clauses in engineering contracts about additional payment benefit both parties. They benefit the contractor because they provide clear grounds for making a claim for additional payment, and a clear obligation on the part of the employer or purchaser to pay. They benefit the employer or purchaser because the list of grounds for additional payment is finite, and because there are usually clear procedures and time limits within which such a claim can be made. This means that the employer gets the benefit of warnings and reasons for the claim, and, at a later date, of particulars supporting the claim.

Not all contracts have the identical grounds for a claim for additional payment, although most of them include breaches by the employer or purchaser of any obligations under the contract, as well as errors in drawings supplied by the employer or purchaser, and errors resulting from incorrect information supplied by the purchaser. Where different forms of contract may differ is in those grounds for additional payment which do not necessarily arise from breach of contract by the purchaser, for example, unexpected site conditions, or special loads. Parties must simply read each contract carefully in relation to such matters.

Calculating the additional costs

If the additional costs are not caused by any breach of contract on the part of the employer or purchaser, then they must be calculated according to whatever formula is provided by the contract, or, if none, at actual cost. Some contracts include an allowance for profit at an agreed percentage, in addition to the actual cost.

If the additional costs are caused by a breach of contract on the part of the employer or purchaser, then again, the usual basis for calculating the sums due is the formula provided for by the contract, but if none is provided, then the basis is that of the law relating to damages for breach of contract. This has, over the years, given rise to a body of case law, mainly arising from building contracts, particularly JCT contracts, but much of this case law would be applicable, in the absence of any different agreed formula, to engineering contracts. In itemizing a bill for additional costs, let us take an example of a mistake in a drawing supplied by the engineer, which might mean that certain metal components had been wrongly machined. Following an abortive attempt at fitting them, it has become clear that they have to be returned to the works to be re-machined. Clearly, the cost of re-machining is an allowable additional cost. But it is only one part of the full bill of costs, which might include transport, fitting, dismantling, money paid to other parties (such as sub-contractors due to the delay), and so on. If a delay is prolonged, this means that the overheads of the contractor are prolonged, and in an appropriate case the contractor might make a claim on such grounds. If the error is discovered during a period of high inflation, the prologation may give rise to increased costs on that ground. In certain circumstances it has been held that interest or financing charges are payable to the contractor, but this would only arise where it is clear to the employer that, for example, the contractor is having to hire, lease, hire-purchase or otherwise obtain finance for a particular piece of equipment, and where it is clear that the charges are increased by the prolongation.

A full bill of costs for a major breach of contract, or error by the employer or his engineer on his behalf, might include any or all of the following heads of claim, provided of course that the contractor can substantiate each head of claim.

1 The actual cost of carrying out additional work, such as the re-machining of an item.
2 Increased costs, if any.

3 Transport costs.
4 Wages and stores.
5 Interest or financing charges, if actually incurred due to the breach by the employer.
6 Accommodation, plant, tools, storage.
7 Money paid by the contractor to other parties, such as to sub-contractors. This arose in the case of *Croudace Construction Ltd* v. *Cawoods Concrete Products Ltd* (1978), in which it was held that a contractor could claim for extra payments which the contractor had had to make to sub-contractors who had had to spend additional time on site due to a delay.
8 Management costs and superintendence. This is calculated on similar principles to the previous heading, but it must be emphasized that the employer is entitled to call for proper records to substantiate a claim. The courts have disallowed a claim for overheads which was calculated purely as a percentage of other items: *Tate & Lyle Food and Distribution Ltd* v. *Greater London Council* (1982).

Whether or not an allowance for profit can be added to these items depends upon the terms of the contract. If the contract does not actually state that profit may be added, then the question of whether or not the contractor will be entitled to add profit will fall to be determined by the common law. Under common law principles, a claim for profit can only be made if profit has been lost. In the context of an engineering contract, this would arise, for example, if the problem created by the employer caused the contractor necessarily to use resources which the contractor could have been using profitably elsewhere. Whether or not such a claim can be substantiated in any given instance depends entirely upon the individual circumstances of each case.

Suspension and termination of engineering contracts

The starting point for discussion of these issues is the basic principle that for either party to an engineering contract to suspend performance of it, or to terminate any part of it, is a breach of contract. Suspension by the contractor involves breach, because it means that there would be a failure to comply with the programme, and probably a failure to meet the date for completion. For the employer or purchaser to suspend performance involves breach, because it hinders the contractor in the performance of his duties, and very likely involves extra costs to the contractor. Termination on either side is, in principle, a breach of contract, because the party terminating is showing an intention not to carry out his part of the bargain.

However, these basic principles need to be heavily qualified, since most modern engineering contracts contain complex provisions about suspension and termination, and in many cases give to either one or both of the parties the *right* to suspend or to terminate, in certain circumstances.

The most likely example of the right to suspend performance of the contract is where such a power is given to the engineer by the terms of the contract. If such a power is given, contractors should ensure that the contract also contains provisions for additional costs to be paid to the contractor. Good contracts will also recognize that a contractor's cash flow could be seriously affected by suspension of important works and of the delivery of expensive plant. Provisions which entitle the contractor to payment for plant which has been affected for longer than an agreed time by the suspension, will greatly assist the contractor. Provisions which entitle the contractor to treat the contract as ended due to the employer's default, if suspension continues for longer than an agreed maximum period, will be needed by the contractor as a last resort, if the suspension of works becomes a matter of serious commercial inconvenience to the contractor. It cannot be stated too strongly that clear provisions on all of these points should be identified by the contractor *before* the contract is made. Good examples appear in the *MF/1 General Conditions of Contract* (1988).

Suspension and force majeure

We have already noted that *force majeure* is usually provided for in a contract in such a way as to excuse whichever party is affected by *force majeure* (which may in certain cases be both parties) from performance of contractual duties until the circumstances constituting *force majeure* cease to exist. There can, of course, be suspension of work under an engineering contract without the existence of *force majeure*. In such circumstances there should be little difficulty in applying the provisions of the clause dealing with suspensation and additional payment. Equally, there can be circumstances constituting *force majeure* without any suspension of work being done under the contract: the *force majeure* circumstances may give rise to delay, but not necessarily to an order to suspend work.

However, it is possible that a situation of *force majeure* may be so serious that an instruction to suspend work has to be given. In such a case a problem arises as to whether the contractor is entitled to additional payment. If the normal rules as to *force majeure* apply, the answer is no. On the other hand, if there is a clause entitling the contractor to claim additional costs arising as a result of an instruction by the engineer to suspend the carrying out of the works, then this may entitle the contractor to make a claim under this clause. Everything depends upon the exact wording of such a clause, and its relationship to the *force majeure* clause.

Termination

One or more clauses dealing with the subject of termination are an important part of an engineering contract, and we have looked at such clauses already in this work in Chapter 3. What needs to be dealt with here are the types of provisions which may exist, and the grounds that may be stated in such provisions for termination. As this work aims to touch on all types of engineering contracts, including those written solely for the benefit of one of the parties to such a contract, we need to be aware that *not all engineering contracts contain reciprocal provisions* for termination. Many engineering contracts, particularly those drafted by purchasers, contain only the express grounds for termination by one of the parties. In such cases, this fact does not entirely deprive the other party of the right to terminate the contract, but it does leave that party to rely upon *implied terms*, or upon the common law relating to breach of contract. As this puts a party at a disadvantage, it is normal to find in forms of contract drafted by institutions, clauses entitling both the contractor and the employer to terminate the contract in different circumstances. Even in institutional contracts, it should be noted, these provisions are not necessarily exactly reciprocal. Nor are they identical in different forms of such contracts. A list of possible grounds for termination of an engineering contract is shown in Table 5.2, but it is not exhaustive, and each particular contract

Table 5.2 *Grounds for termination of an engineering contract*

Typical grounds for termination by purchaser (contractor's default)	Typical grounds for termination by contractor (purchaser's default)
1 Insolvency of the contractor	1 Insolvency of the purchaser
2 Failure to execute the work in accordance with the contract; failure to proceed with due diligence; neglect in carrying out obligations under the contract	2 Failure of the purchaser to make payment within the time stated by the contract
3 Abandoning the contract	3 Obstructing or interfering with the issue of any certificate by the engineer
4 Assigning the contract	4 Replacing the engineer against the reasonable objections of the contractor
5 Suspending the execution of work.	

must be looked at carefully to see precisely which grounds for termination have been stated. Apart from any express formulation of grounds for termination, general principles of the law of contract apply, and termination would be possible by one party on grounds of a serious breach or a repudiation of obligations by the other party.

Some legal questions answered

Is it essential to define or to provide a list in the contract of circumstances constituting force majeure?

No, it is not strictly necessary. Simple or minor works contracts can work perfectly well if they refer simply to 'circumstances beyond the reasonable control of the parties'. However, in a larger engineering contract it is better to have some form of definition, since there are well-known areas of difficulty, such as whether or not, and to what extent, strikes and other industrial disputes amount to *force majeure*, and whether or not shortage of materials may constitute *force majeure*. A good contract will definitely rule such matters as being either within or without the definition.

Is it true to say that a list of circumstances amounting to force majeure can limit the ability of a party to make a claim under the clause?

Yes, it can, although for this reason many contracts contain phrases such as, 'including but not limited to the following circumstances', followed by a list of the relevant circumstances. Without words which keep the list open-ended, it could be interpreted a being finite. Further, there is a rule of construction of documents known by its Latin name of the *eiusdem generis* rule, that is, the rule about 'things of like kind'. This rule means that even if you try to give an open-ended list, the extrapolation of it will be presumed to be a continuation of things *similar in nature to the items listed*. Bearing in mind this rule, it would be a mistake to write a *force majeure* clause stating that *force majeure* 'includes earthquake, flood, adverse weather conditions and any other cause beyond the parties' reasonable control'. Applying the *eiusdem generis* rule, such a clause, which appears to be open-ended, in reality is limited to natural physical conditions, and it would not be possible to stretch the interpretation of the clause to include, for example, a strike or civil commotion, or failure of a public utility.

Is the expression 'Act of God' equivalent to force majeure, and is it advisable to use this expression?

No, it is not equivalent to *force majeure*. 'Act of God' is only a collective expression for circumstances of the kind mentioned in the previous paragraph, that is, weather conditions or natural physical disasters not caused by any human agency. Most parties will want their *force majeure* clauses to be far wider than this, and to include various forms of war or hostilities, as well as riot and civil commotion, and acts of Government, and in many cases industrial action.

It is difficult to advise as to whether or not the expression 'Act of God' should be used in an engineering contract. In the past it has been popular, because of the width of the expression, but it is now becoming less frequently used in modern contracts, perhaps because of its lack of precision, and perhaps in recognition that the expression does not have the same validity in every culture.

Have claims for extension of time on grounds of force majeure ever been known to fail because of failure to comply with the formal requirements?

Yes, they have, although this does not mean that the formal requirements are always strict, or that they are always strictly enforced. There is, however, usually a requirement in the contract that prompt and accurate notice should be given by the party affected by the circumstances to the other party, with supporting details as soon as they are available. If this formal requirement is not carried out, the other party may, at a later date, dispute the accuracy or even the reality of the claim, particularly if it is made at a much later stage and if it has financial implications: the financial implications are usually that the contractor claiming the benefit of the *force majeure* clause is seeking to be released from liquidated damages which the purchaser intends to apply at the conclusion of the contract.

An example is given by the case of *Intertradex SA v. Lesieur-Tourteaux SARL* (1978). In this case there was a relatively generous *force majeure* clause in the contract, which provided:

> *Force majeure*, strikes, etc.: Sellers shall not be responsible for delay in shipment of the goods or any part thereof occasioned by any breakdown of machinery or any cause comprehended in the term *force majeure*. If delay in shipment is likely to occur for any of the above reasons, Sellers shall give notice to their buyers. ... The notice shall state the reason(s) for the anticipated delay.

There was a delay, for which the buyers were seeking damages, and the sellers claimed that the delay was caused by *force majeure*. This claim was disputed, because the notice was inaccurate, and had stated that the entire period of delay was due to one cause, namely the breakdown of an electrical distribution panel, which would have been covered by the particular *force majeure* clause, while the reality was that this was only one of the reasons for the delay, the delay being also due to other causes which were not covered by the *force majeure* clause. The case went to arbitration, and subsequently to the Court of Appeal, where it was held that the notice of *force majeure* was bad. This did not entirely rule out the claim of the sellers, but it meant that they would have to give a revised notice and to prove to the arbitrator that the electrical breakdown would have been sufficient on its own to account for the full period of delay.

Is the breach by the employer or purchaser of any of his general obligations under the contract a ground for termination by the contractor?

As has already been noted in this chapter, termination clauses are seldom exactly reciprocal and need not necessarily be reciprocal at all. While it is common to find that a contract will provide that the purchaser has the right to terminate the contract on the ground that the contractor is in breach of his obligations under the contract, there is often no express equivalent right of the contractor. The reasons for this are partly the fact that the majority of engineering contracts are written for the benefit of purchasers, and partly the fact that it is seldom that a contractor will actually wish to terminate an engineering contract, thereby losing the commercial benefit of that contract, except on the grounds already mentioned, namely those of non-payment and insolvency on the part of the purchaser. However, instances have arisen where a purchaser has seriously failed to meet his obligations under an engineering contract, such as failing to make the site available so that the contractor may proceed with the works according to the programme and completion date originally envisaged. In such circumstances, as long as the contract is explicit about this being a purchaser's obligation, and as long as there is a stated date or period within which commencement should take place, there is little doubt that the contractor can claim for his additional costs due to the delay. However, it may be that the delay is so serious that the contractor wishes to terminate the contract so as to proceed more effectively with other, less problematical work. If this is the case, and if there is no contract provision which states that the contractor is entitled to terminate on this ground, then the contractor will have to rely upon general principles of the law of contract. The most likely solution for the contractor will be to claim that the purchaser has *repudiated* his obligation under the contract, or is in serious breach of a material obligation under the contract, and on such grounds the contractor will, if need be, be entitled to terminate.

What is the financial position after the termination of an engineering contract?

This depends, firstly, upon what the contract says. Secondly, it depends upon which of the parties has elected to terminate the contract and on what grounds. One cannot generalize, because the circumstances are too diverse and potentially too complex. For example, if the contractor were to become insolvent, the purchaser may wish to terminate the contract, but the contract might also give the purchaser the option of asking the receiver or liquidator or administrator to complete the work under the contract. The receiver, etc., would not be compelled to complete the work, but might agree to do so.

Engineering contracts may provide for the contractor to be paid for all work up to the date of termination, but in such cases there are usually provisions for deduction or recovery of any additional costs which the purchaser has incurred due to the fact of termination and the need to have the work completed by a different contractor.

Is force majeure ever a ground for termination of an engineering contract?

A well-written *force majeure* clause should contain a provision providing that either of the parties may terminate the contract if *force majeure* lasts for a continuous period exceeding a specified period of time. The law of frustration, under common law, would provide that a contract is frustrated if performance is so seriously disrupted as to become radically different from anything the parties had originally envisaged. However, the common law of frustration is too vague to be relied upon, and it is better to have a precise agreed maximum period after which either or both of the parties may elect to terminate. If no such period is provided for, it is possible that the *force majeure* clause might even be interpreted as permitting an indefinite period of delay in a situation where *force majeure* circumstances have occurred. A typical provision in an engineering contract might be one providing for possible termination at the option of either party after 120 days, with agreed subsequent financial arrangements, such as payment for works done up to the date of termination, together with certain agreed costs. Such a clause would be both desirable and reasonable from the point of both parties.

What is the legal position if progress in an engineering contract is held up by third parties?

This is another potentially complicated issue, since a great deal will depend upon who the third parties are. If they are sub-contractors to a main contractor, then any delay or default by those sub-contractors will be the liability of the main contractor, whose recourse will be to try to recoup his loss against the offending sub-contractor. A complication is that some contracts, such as the JCT, and the I.Chem.E. (1981 up to 1993), permit the contractor to claim for an extension of time due to delay on the part of a *nominated* sub-contractor, which the main contractor has made all due effort to avoid or reduce, by proper supervision, etc. By no means all contracts, or even all editions of particular contracts, contain this provision, and it should be carefully checked in every case. If the third party causing a delay or disruption is employed directly by the purchaser, then any consequences of the delay or disruption will be accountable to the purchaser.

Is it possible to summarize the key provisions of an engineering contract relating to progress?

All engineering contracts are different, and will not necessarily contain or even require identical provisions. However, a good engineering contract should contain as many as possible of the following list, which is not intended to be exhaustive.

1 A time or date for the purchaser to give to the contractor possession of, or access to the site, or facilities as agreed for carrying out the work.
2 A requirement that the contractor submit a programme for carrying out and completing the contract work, and a time or date for the submission of that programme.
3 A requirement that the engineer approve (or disapprove) the programme, and a time within which this should be done.
4 Times or dates for the provision by the contractor of drawings and other required information.
5 Times or dates for the approval (or disapproval or amendment) by the engineer of drawings and other data submitted by the contractor.
6 Times or dates for the purchaser to supply to the contractor any drawings and other information which it is the purchaser's obligation to supply; similar provisions for free-issue of goods.
7 Times or dates for tests.

8 Clear payment milestones.
9 A time or date for completion.
10 A *force majeure* clause.
11 The rights of the contractor to additional payment on account of delay or suspension.
12 The rights of each party to terminate the contract.

6 Quality and fitness for purpose

The legal and commercial kaleidoscope

This chapter deals with the prime objective of an engineering contract, which is to obtain equipment, plant or works which are as specified, fit and free from defects, and perform to the required standards. This is not simply a matter of legal content and documentation in the relevant contract, nor is it simply a matter of commercial and technical expertise. It is a combination of all of these things: a legal and commercial kaleidoscope in which a number of ideas intersect and interrelate. The aim of this chapter is to explain and clarify the issues, so as to show where responsibilities lie, if problems should occur. It is not an easy task, as the law does not stand still for very long in this area, and, further, each case presents its own peculiar set of circumstances.

Setting the commercial scene

In the matter of quality and fitness, legal action should be a last resort, although legal issues should always be kept in mind throughout the negotiation of the contract. The conditions of contract are the main instrument for providing the legal framework for these matters. However, it is best, initially, the approach things from a commercial and technical point of view. If this is done properly, then legal considerations are in the overwhelming majority of instances a 'spare wheel': something to be kept in hand and to be used only as and when necessary.

The commercial scene is set by both parties, but particularly the purchaser, addressing the following issues, and making sound decisions which are then properly recorded and, where required, reflected in the contract. These are:

1 *The technical feasibility of the project.* Will the particular idea work? Is it technically possible? What are the risks? Is the purchaser relying upon his own expertise and judgment, or will a contractor undertake responsibility for the design and feasibility?

2 *The commercial viability of the project.* What will the financial returns be? Is a contractor prepared to offer performance guarantees? Have all the factors of use been taken into account; such as, location, operating shifts, servicing time, availability of spares, variations in application, etc.?
3 *Choice of a competent and reliable contractor.* This is the crucial commercial and technical choice, and legal protection can only partly make up for a bad decision in this area.
4 *Clarity of the specification.* Both parties may take some responsibility for the specification, or it may be the responsibility of one party only. If the specification is not clear, the responsibilities are not always easy to determine.
5 *Clarity of the responsibilities of the parties.* This point goes hand with the previous point, but also relates to matters such as the condition of a site, the facilities available, the provision of information and data, the provision and/or approval of drawings, samples, models, etc., and the provision of 'free-issued' goods and materials.
6 *A good system for the elimination of risks and for the early identification of defects and deficiencies.* This is a point of good project management. If a problem or discrepancy can be discovered at an early stage, such as during a pre-delivery test or inspection, it will be easier and cheaper to put it right, and less inconvenient for both parties. The schedule or system of tests and inspections must be properly reflected in the contract.
7 *Financial leverage through the system of payment and/or security.* If commercial contracts were put purely upon the basis of either trust, or long-standing relationships, or confidence in one's legal rights, this aspect of engineering contracts would not be necessary. However, none of these things is entirely sound or reliable as a means of securing good performance, so it is common in engineering contracts for the purchaser to require financial leverage. This is partly achieved by control over the system of payment to the contractor: money may, under the terms of the contract, if so desired, be released only against proven value or achievement: a system often known under the term of 'milestones'. Additionally, the purchaser may decide that he requires the security of money kept back for a period of time after delivery and taking over: 'retention money'. In larger contracts, particularly overseas contracts, the purchaser will often want the further security of a 'performance bond'.
8 *Commercial leverage.* The purchaser of goods or services should always consider carefully the commercial incentives to the contractor in relation to compliance, performance and aftersales service. Problems can arise once equipment has been put to use. Financial leverage may have become exhausted with the passage of time. Legal remedies are

slow and uncertain. If the contractor has a commercial incentive, such as an interest in continuing the business relationship, then this can be of value to the purchaser. Conversely, in some relationships, particularly when a main contractor is purchasing from a nominated sub-contractor, there may be a damaging lack of incentive on the part of the sub-contractor.

Some examples and case histories

Cammell Laird and Co. Ltd v. Manganese Bronze and Brass Co. Ltd (1934)

This case illustrates the importance of allocating responsibilities clearly. The purchaser was the contractor responsible for the construction of a ship. The seller was the manufacturer of the ship's propellers, and was a specialist in this field. The drawings were supplied by the purchaser, and the contract was to comply with the drawings. Normally, this would have placed responsibility upon the purchaser if the drawings were based upon a miscalculation. However, in this case, the specification stated that the leading edges of the propellers were not accurately shown in the drawings, and that it was for the sellers to use their own judgment as to how to taper the propellers to 'fine lines'. In the event, the ship failed a performance test, because one of the propellers created noise. If the noise had been due to any part of the design which was supplied by the purchaser, the purchaser would have had no claim against the seller. However, once it was established that the cause of the noise was the finishing and shaping of the leading edge of the propeller, it was held by the court that the seller was liable (the propeller was unfit for its purpose within the meaning of section 14 of the Sale of Goods Act 1893.)

Aswan Engineering Establishment Co. v. Lupdine Ltd (Thurgar Bolle Ltd, third party) (1987)

This case illustrates a failure to examine the technical feasibility of a project in all its details, as well as the failure of the purchaser to place full reliance upon the seller, as regards the correct specification. Plastic containers were supplied to Lupdine Ltd by Thurgar Bolle Ltd, for the purposes of transporting liquid waterproofing compound to Kuwait, to be used by Aswan Engineering Establishment Co. Apparently, the fact that the containers were to be stacked in larger metal containers, with exposure to very high temperatures created by the natural climate of Kuwait, was not made known to the sellers, Thurgar Bolle Ltd. Further, it appears that no consideration had been given to the feasibility or otherwise of stacking each loaded plastic container on top of another, in piles, causing each container to bear a much heavier load than the weight of its own contents. The evidence was that if the matter had been fully discussed before the contract was made, a different method of stacking, on slats or in metal frames, would have been recommended.

In the event, the plastic containers collapsed when stacked in Kuwait, and their contents, some 35 100 kg of compound, were lost. Lupdine Ltd had to defend a claim for breach of contract brought by Aswan Engineering Establishment Co., and in turn, Lupdine Ltd brought a claim for breach of contract against Thurgar Bolle Ltd. This claim was on the grounds that the plastic containers were not of merchantable quality, and that they were not fit for the purposes for which they were bought. Lupdine's claim failed on both grounds. The containers were exactly as specified, and were not in any way defective. They were appropriate for all normal uses. Consequently they were of merchantable quality. As to the argument that they were not fit for the purpose for which they were bought, Lupdine could only win this point if Lupdine Ltd could show that the seller was aware of those purposes (that is, the location, the temperature and the method of stacking), *and* that the seller had been relied upon to ensure that the specification of the containers was suitable for those purposes. This was what Lupdine could not show.

IBA v. EMI, IBA v. BICC (1980)

This case illustrates how important it is, from a contractor's point of view, not to take on a responsibility for design and suitability unless one fully intends to do so and is fully aware of the implications. The contractors, EMI, had contracted with the ITA (later changed to the IBA) for a project involving the *design*, supply and delivery of a television mast, together with other services, at Emley Moor, in Yorkshire. When EMI tendered for the contract, the instructions were to incorporate into the tender the design of the nominated sub-contractor BICC Ltd. This design had been prepared at the request of the ITA, and had been sent to it by BICC Ltd.

After being erected, the mast was operational for a while, and then collapsed due to a combination of wind and icing, which created stresses known as 'vortex shedding'. In the claim that followed, the House of Lords held that on the facts, EMI had accepted responsibility for the design. Lord Scarman stated that in the absence of a term negativing the obligation, if one contracts to design an item for a purpose of which one is aware, one undertakes that the item is designed so as to be reasonably fit for that purpose. EMI was therefore liable to the IBA. It should be added, so as to avoid any misunderstanding, that BICC was liable in turn to EMI, and BICC was also liable directly to the IBA for negligence in giving assurances as to the fitness of the design.

Ruxley Electronics and Construction Co. Ltd v. Forsyth (1995)

That a case involving no more than £21 000 should go all the way to the House of Lords is, in itself, surprising. The point of principle at stake is, however, of potential importance to the whole of the construction industry, and one can only speculate as to how far reaching the decision will be. The essential issue from the point of view of the purchaser, is whether or not such damages as may be awarded by a court in respect of defective design or work are likely to compensate the purchaser in full for the defects. Until the Ruxley decision, it was thought that by and large, the answer was in the affirmative. If work was defective, damages, until this case was finally decided, usually reflected the cost of reinstatement.

Mr Forsyth contracted to purchase a swimming pool, to be constructed by Ruxley Electronics and Construction Co. Ltd. On completion, it turned out that the deep end was 9 in. shallower than was specified. Mr Forsyth contended that he should be awarded the cost of having the pool re-dug, until the deep end was the correct depth. The cost of this was £21 000. To Mr Forsyth, the loss suffered by the failure to meet the specification was substantial, since the lack of depth made diving into the pool somewhat hazardous.

In normal commercial contracts, in such circumstances, the two alternative ways of calculating damages are either to assess the diminution in value of the property, or to assess the cost of reinstatement. The first of these methods of assessment was not, however, applicable in this case, since the pool had to be valued with the whole house and garden, and there was apparently no real diminution in value. The Court of Appeal awarded the cost of reinstatement of the pool. However, the House of Lords reversed this decision and held that the cost of reinstatement was disproportionate to the breach, and instead awarded the sum of £2 500 for 'loss of amenity'.

The case illustrates perfectly the fact that care in the commercial and technical and management aspects of an engineering contract is of great importance, since the law may not provide adequate compensation for all defects. The potential impact of the case on construction contracts is, however, a matter for speculation. It may be that the decision will be confined to its domestic circumstances, as it turns upon the apparent injustice of an award of £21 000, and the apparent ability of the court to assess damages for loss of amenity. These considerations are less likely to apply in a non-domestic commercial contract. Nevertheless, the so-called 'Ruxley defence' may have great appeal to claims consultants and lawyers acting for contractors in commercial cases.

Where are the terms of a contract concerning quality and fitness for purpose to be found?

The answer to this question is that the terms and their full implications are to be found in a number of places, together with a certain degree of overlap and interaction, and it is this fact that makes this area of the law somewhat complex and kaleidoscopic. In order to simplify matters, it is best to divide the terms into two categories. These are: *'express terms'* (these are in the contract documents, except where the contract is made orally); and *'implied terms'* (these are to be found in the common law and in a number of Acts of Parliament. They are often referred to as 'statutory rights'). Having made this distinction, it will usually be found that the express terms appear in the *specification*, as well as in terms of the contract relating to *design* and *conformity with standards*.

Implied terms can be found in the Sale of Goods Act 1979, as amended by the Sale and Supply of Goods Act 1994, and the Supply of Goods and Services Act 1982. The latter Act provides implied terms in commercial contracts which are not for the sale of goods, but for hire, construction, repair or servicing. The chart that appears in Figure 6.1, although not

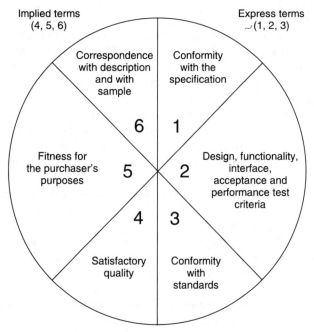

Figure 6.1 *Comparison and relationship of express and implied terms as to quality in an engineering contract*

exhaustive, provides a simple illustration of the relationship between express and implied terms in an engineering contract.

As there is a certain amount of overlap between the express terms and the implied terms, it may be wondered why the implied terms should exist at all. The answer to this is that they serve to provide minimum standards of performance of commercial contracts, and in particular they exist even where there are no express terms. Express terms may be lacking, or incomplete, in contracts which are made informally or in an emergency. It may also be the case that with even the most detailed contracts, there is some matter which is not fully worked out in the express terms, and which may be amplified or explained by the implied terms. Briefly, and without attempting to reproduce the precise wording of the relevant statutes, the implied terms are as follows:

Correspondence with description

This term is only implied where there is a sale of goods by description or an agreement to transfer goods by description. The fact that the legislation states that the term implied is a *condition* of the contract means that it is an important term of the contract that goods should correspond to description.

Correspondence with sample

This implied term is similar to the one about description. It is important to note that if a sale is by *both* sample and description, it is not sufficient that goods should correspond with sample only: they must also correspond with description.

Satisfactory quality

For many years this was known as the implied term as to 'merchantable' quality. Broadly speaking, it is the implied term that goods will be of an acceptable standard of quality, fit and free from defects or damage. The problem with the word 'merchantable' was that it was no longer part of everyday language, and it required a mass of case law to give it meaning. A great deal of the case law was either contradictory or was difficult to extract any meaning from, so it was recommended that the expression should be changed to 'satisfactory quality'. This was done by the Sale and Supply of Goods Act 1994. Whether the change will prove to be more than merely cosmetic, only time will tell, but at least the amendment to the law contains a fresh attempt at definition. It is confirmed that 'satisfactory' takes account of the state and condition of goods, their appearance and finish, and freedom from minor as well as major defects, their safety and their durability.

96 Quality and fitness for purpose

Fitness for purpose:

This has probably been, and remains, the most controversial of the implied terms, perhaps because of the inherent difficulty of interpreting the purpose for which the goods were bought or acquired. With some goods the purpose is simple and self-evident, and does not differ from case to case. The seller will be assumed to be aware of the purpose, and it will be easy to imply the term about fitness for purpose. In other cases, the goods will have a number of possible applications, and the buyer may attempt to stretch the application of the goods beyond normal possibilities or, as in the *Lupdine* case already mentioned in this chapter, to use the goods in abnormal circumstances. Is the seller liable for the failure of goods to fulfil the buyer's purposes in such cases? Clearly not, since the seller is not aware of the buyer's purposes and it is not reasonable, in the circumstances, for the buyer to rely upon the skill and judgment of the seller. The *Lupdine* case, and the case of *IBA* v. *EMI*, already discussed, show contrasting examples of when it is and when it is not reasonable for the buyer to rely upon the seller to ensure that an item is fit for a particular purpose.

Defects liability and express warranties

Not every engineering contract or contract for the sale of goods contains an express warranty as to the quality of work, goods or materials. This may be because a small order simply relies upon the specification, and any defects are taken care of either by commercial means or by use of the implied terms. Or it may be a deliberate decision on the part of the purchaser: there have been instances where the 'warranty' offered by a seller has been so inadequate (although the goods have in other respects been competitive) that the purchaser has declined to accept it, and has preferred to rely upon the implied terms or 'statutory rights'. To evaluate such a decision, and to be able to assess the relationship between an express warranty, we must consider the subject of defects liability generally, and then examine the ways in which an express warranty will affect the position of the parties.

Common law liability of seller in respect of defects

If goods are delivered or if work is done under an engineering contract, and if there is a clear breach of any of the purchaser's statutory rights, then this, under common law and under statute, is what is known as a breach of 'condition'. This means that the breach is classified as being serious enough to entitle the purchaser to reject the goods or work. The result will be that the purchaser will be entitled to withhold the whole price, if not yet paid, or to claim a refund of the price if it has already been paid. There is, as a matter of common law, no rule that a purchaser must accept replacement goods or work, and where purchasers do decide to take the replacement of goods route, this is purely a matter of practice or convenience.

The Sale and Supply of Goods Act 1994 introduced a change in the law as regards sales of goods to commercial customers (that is, to buyers who are not consumers). This change means that a business buyer will not be entitled to reject goods if the breach 'is so slight that it would be unreasonable for him to reject'. This amendment to the law can hardly be said to have brought any more certainty to the law, as its terminology is about as easy to measure as the proverbial piece of string. Presumably what it will do is encourage more buyers to settle with sellers on reasonable terms. The buyer does not lose his normal common law right to claim damages in respect of defects, so where the defect falls into the 'slight defect' category, the buyer will have the right to pay less than the full price.

As regards work to be done under an engineering contract, the purchaser is entitled under his statutory rights to refuse to accept or pay for work which is not of the quality required under the contract or which is not fit for the purpose specified. If the contract is an entire contract, that is one which does not have specific 'milestones' for payment, then the purchaser is entitled to withhold the entire contract price if the work is defective or unfit. There is a common law equivalent to the 'slight defects' rule discussed in the previous paragraph, and this is known as the rule of 'substantial' performance. What this means is that if the defects are such that the work or structure has serious deficiencies, or is not capable of being put to the use intended, then the contract has not been substantially performed, and the purchaser can withhold the price. If, however, the defects are minor and capable of being corrected at a later stage, and if the work or structure can be put to reasonable use, then the purchaser is not entitled to withhold payment, although deductions can be made if the defects are not later on put right.

In the case of *Bolton* v. *Mahadeva* (1972), a plumbing and central heating engineering contractor made a contract to design and supply a central heating system. The system was installed, but when tested it was found to have serious defects. It gave out fumes when switched on and produced inadequate heat, due to insufficient radiators and inadequate insulation. The purchaser withheld the price and the contractor sued for payment. On appeal, the Court of Appeal held that payment was only due when the contract had been substantially performed, and in this case the discrepancies were such that substantial performance could not be said to have taken place. To rectify the defects would have cost the equivalent of one third of the contract price, so the defects could not be called 'slight defects'. As there were no provisions in this particular contract for stage payments, it was an entire contract, and the purchaser was entitled to withhold the entire price.

Time limits for rejection of goods

A purchaser may not wait indefinitely to reject goods; this is because it is not reasonable to expect a seller to take back goods which have been used or stored beyond a reasonably short period of time. The law does not specify any particular period, because this would not be possible. The law simply gives the broad principle, and then leaves it as a matter to be assessed from case to case. It is, however, open to the parties to specify in the contract a period within which goods may be rejected. Such a period would, as a matter of law, have to satisfy the test of reasonableness.

Damages

Any breach of a term of a contract in theory entitles the party not in breach to claim damages, or to deduct damages from any unpaid part of the price. If goods or engineering work are defective, the purchaser may have a claim for damages, for several possible reasons. One reason, which has already been examined earlier, is that delays may result from goods or work being defective and having to be rejected. Another possible reason is that the purchaser may decide, in the circumstances, not to reject the defective goods or work, but to accept it and to pay less. The deduction from the price (to be assessed from case to case) is a possibility both under common law and under statute: Sale of Goods Act 1979, section 53. Claims for a partial refund of money already paid can also be made in such circumstances.

A further point that must be made is that damages due to defective goods being delivered or due to work being defectively performed may be in excess of the purchase price. The principle of indemnity applies here, and the buyer is entitled to be indemnified against any loss arising naturally in the usual course of events from the breach, or against any other loss which the parties would have had in mind at the time of the making of the contract. Attempts to make this into an exact science tend to be somewhat frustrating, and exceptional cases (such as the *Ruxley* case looked at earlier in this chapter) are capable of producing unusual arguments and unusual results. It can, however, be said with some degree of safety that there need be no relationship between the contract price of goods and the damages that can be claimed if they are unfit or defective.

In the case of *Harbutt's Plasticine Ltd* v. *Wayne Tank and Pump Co. Ltd* (1970), new plant and equipment was installed for the purchaser, a plasticine manufacturer. The premises were an old mill, which had been converted into a factory. The design and choice of materials proved to be defective, and in particular there had been a failure fully to take into account the temperature of heated and liquified material which it was intended to pump into the factory to be turned into plasticine. A fire broke out and destroyed the mill. Clearly the case was one of a design and choice of materials being unfit for the specified purpose. The measure of damages was one of the main issues in contention, and it was held by the Court of Appeal that the owners were entitled to damages which reflected, among other things, the cost of rebuilding the factory. The argument for the defence was that this would far exceed the value of the old factory, since the design would be more modern and there would be 'betterment'. However, the court held that the owners had no option but to replace the factory, and the design, although to modern requirements, did no more than replace the old factory.

This 'new for old' principle has a weight of precedent to support it, although it occasionally appears to surprise insurers and loss adjusters. The misunderstanding arises from failure to keep distinct two quite separate issues. The first issue is that if you are the client of an insurance company and have insured goods, the basis of insurance depends upon the contract of insurance, and may in some cases be 'new for old', and in other cases may be replacement that takes account of age, wear and tear and depreciation. The premium paid will reflect the difference between the two types of insurance. However, damages payable by a party in breach of contract, where the breach of contract causes damage to property, do not depend upon the basis of insurance that has been chosen. They depend upon what a court would consider to be a reasonable way of replacing the damaged property. With damaged buildings, this is normally re-instatement. With goods damaged beyond repair, the method normally accepted by the courts is replacement with new goods. In some exceptional cases, replacement with second-hand goods would be the appropriate method; this would be the case where there exists a ready market in which second-hand goods of the kind destroyed or damaged can be obtained. But in most commercial cases, there is no such market, and it is not reasonable to expect a well run and safety-conscious commercial buyer to replace plant and equipment with second-hand equipment.

In the case of *Dominion Mosaics Tile Co. Ltd* v. *Trafalgar Trucking Co. Ltd* (1989) a company whose equipment was destroyed by another party was held to be entitled to replace the equipment by purchase of new equipment at a cost of £65 000, even though the destroyed equipment had only cost approximately one fifth of this sum. The reason for this award was quite simply that the party seeking the damages was doing no more than replace the exact equipment lost, on a new basis. The difference in cost was entirely due to a rise in prices.

Express warranties in respect of work or goods

Having outlined the basic common law and statutory position in relation to goods or work and materials, it is now necessary to take a closer look at the well known concept of the 'warranty', which may in some contracts be called a 'guarantee', or in some contracts may simply be described as 'defects liability'. While such a 'warranty' may, initially, seem a desirable undertaking to have, from the purchaser's point of view, and while in some cases its content and quality may even be influential in helping the purchaser to choose between different types of goods or suppliers, it should be understood from the outset that a 'warranty' in respect of goods and services may well be a double-edged weapon. It may help the purchaser to obtain aftersales service or replacement of defective goods, but it may also help the seller or contractor to limit risks and liabilities under the contract. The precise content and effect of such a warranty depends upon its detailed wording, but in many cases a warranty is, in effect, a form of trade-off, by which the seller or contractor offers rights and remedies to the purchaser which are different to and more convenient than those available under common law and statute. In return for these rights and remedies, the buyer modifies, or even in some cases gives up, the rights and remedies he would normally have by law. The reason for this trade-off is one of commercial realism and acceptance that one party to an engineering contract cannot be expected to bear all the commercial risk which arises as a result of selling goods or designing and building plant and equipment. The position of the private consumer is quite different. One cannot speak in the same way of a division of commercial risk, and consequently the private consumer retains his or her full statutory rights under a contract for the sale of goods or of services. It is not possible or lawful to exclude the private consumer's statutory rights. More detail on this particular point will be given in Chapter 7.

However, with a commercial engineering contract the problem of risk which both parties must confront is that an engineering company with a medium-sized capital and turnover may contract to produce equipment for a company in a field such as brewing, food manufacturing, oil refining, or the production of chemicals, which has a capital and turnover of many times the value of the contractor. If the equipment should fail, after delivery and acceptance, the stoppage may cause serious commercial losses. The 'warranty' or defects liability clause in the contract normally takes some account of the possibility of such commercial losses

Quality and fitness for purpose 101

(often, but not strictly accurately, known as 'consequential loss') and modifies the liability which would fall upon the contractor under usual principles of law. In short, the contractor often excludes or limits such forms of liability, partly as a trade-off for the aftersales response that is offered under the warranty, and partly because it is understood by both parties that the larger company, the purchaser, is in a better position to bear or to mitigate or to take precautions to avoid the commercial losses.

Evaluating a warranty

A warranty of the kind discussed in this chapter may be evaluated by asking and answering a number of questions which relate to a hypothetical failure or defect in the goods or services to be provided under the contract. The list of questions which follows is by no means exhaustive; nor can it be promised that every engineering contract necessarily supplies answers to these questions. However, the careful reader and negotiator of engineering contracts should try to find answers to as many as possible of the following points.

What is the duration of the warranty?

This question should be looked at not only in terms of a given period of time, such as twelve months, but also in relation to the date of commencement of the warranty. In engineering contracts where components are purchased well ahead of likely completion of the project, it may be advisable to require postponed dates for the commencement of warranties, so that warranties do not expire before the works are taken over. Similar considerations apply to items purchased in a chain of supply contracts. The intermediaries between a manufacturer and the end-user do not normally wish to be in a position where the warranties they *receive* expire before the warranties they *give* to their purchasers.

Is the warranty transferable?

This is important for a number of reasons. One is if the party receiving the warranty is an intermediate buyer, intending to sell or to install the items to his own customer. While a contractor will be expected to warrant the goods and services he provides, it may also be of commercial value to be able to pass a warranty on to the end-user directly from the manufacturer. Another reason why the question of transferability is potentially important is that in the case of work carried out on buildings and other premises, warranties are often of considerable duration, so that in the case of a roof, for example, a warranty of ten years would be standard, and warranties of twenty or twenty-five years would be a possibility. After some years the owner of the building may wish to sell it, and if the warranty is transferable, this enhances the value of the building.

What is the coverage of the warranty?

This raises the question of whether the warranty requires the contractor to bear all the incidental expenses of putting right defects or of replacing goods and materials which are not in accordance with the contract. At one time it was fairly common to find warranties under which parts were replaceable at no cost to the purchaser, but labour incidental to the replacing of parts would be chargeable to the purchaser. This practice is less common than it used to be, perhaps because purchasers tend to have a greater part in the drafting of terms of contract than they used to. Transportation of goods which are to be repaired or replaced is a similar issue, and should be mentioned specifically in the warranty, so that it is clear where the cost falls.

What is the scope of the warranty?

This question is different from the preceding one. It is about the distinction between a mechanical and electrical warranty, for example, and a 'performance' warranty. A motor car usually has, at the time of first sale, a mechanical and electrical warranty. It does not usually have a 'performance' warranty,: its fuel consumption, or oil consumption, or the durability of its brakes or clutch, to give examples, are not usually the subject of any warranty. Of course, if the consumption or durability of parts were to fall a long way below expectations, then it might be possible to infer the existence of a mechanical or electrical defect. However, the fine details of performance, in relation to matters such as consumption or durability (or output, in the case of factory equipment) are not dealt with in an ordinary mechanical or electrical warranty, whereas they can be the subject of a detailed performance warranty. With a performance warranty, the parameters of performance, the permitted tolerances, the conditions of use and measurement, and the remedial issues, if specified performance figures are not attained, can all be dealt with.

What is the response time and what are the response methods?

The answers to these questions concern the minute details of a warranty, without which a warranty tells little about how it operates in practice. A warranty under which a response to a reported defect or breakdown can be guaranteed within, for example, twenty-four hours, is of far greater value than one where no such period is specified. The method of response is also important, since a guaranteed response at the purchaser's premises is, again, of considerably more value than warranties under which the purchaser has to return the defective equipment to the supplier.

Are there any geographical limits to the warranty?

The relevance of this question is self-evident, particularly in view of the previous paragraph.

Are there any conditions about use in the warranty?

It is not uncommon for a warranty to contain a general proviso to the effect that the warranty only applies to defects or breakdowns which arise under normal or specified conditions of use. There is nothing particularly wrong with such a statement, and it is of some protection to the contractor. Purchasers should make sure that such conditions about use are not too restrictive.

Are there any conditions about maintenance or servicing?

This is similar to the previous question, and again, it is reasonable for the contractor to seek some protection against the warranty claim which might arise due to failure by the purchaser properly to maintain or service the equipment. The purchaser should make sure that the warranty does not depend upon unreasonable servicing requirements.

What is the relationship of the warranty to the statutory rights of the purchaser?

The statutory rights of a purchaser have already been examined in this chapter. These rights would apply even if there were no express warranty, and in many instances, a purchaser does not have or need an express warranty. The benefit of a warranty is in its detail, and in the fact that it may offer the purchaser additional benefits, other than those which arise out of common law or statute. If the buyer is a consumer, a warranty must be such as not to affect the statutory rights of the consumer. This is a legal requirement. However, the position if the purchaser is a business is more complicated. It is theoretically possible for a seller or contractor to write an express warranty in such a way that it *replaces and excludes* all other contractual rights of the purchaser, whether arising under common law or under statute. However, in practice the seller should only be able to achieve this if the warranty which is offered is one which adequately compensates the purchaser for the rights and remedies which it replaces.

This is partly because the purchaser will not normally contract to purchase goods or work or materials on onerous terms. It is also because in United Kingdom contracts, even a business purchaser enjoys some legal protection against onerous exclusions or limits of liability, and may challenge an unfair warranty as being one which fails to satisfy the test of reasonableness. (Unfortunately the purchaser does not enjoy this protection in international supply contracts, and must therefore be particularly vigilant about the terms of the warranty.)

In the case of *Rees Hough Ltd* v. *Redland Reinforced Plastics Ltd* (1984), there was an order for pipes for a contract made with a water board. The pipes proved to be not of merchantable quality, since they were unfit for normal use of the kind specified, and had cracked in the course of being laid. The conditions of contract contained a term under which the seller accepted defects liability notified within three months of delivery. The term went on the state that all other terms and conditions and liabilities under statute or otherwise were excluded. A complaint in respect of defects was made by the purchaser outside the three-month period stated, and when the matter went to court, it was held that this particular 'warranty', together with the exclusions of other liabilities, did not satisfy the test of reasonableness. The result would presumably have been different if the stated period had been measured in years rather than in months.

Is the warranty extended during periods of non-use due to defects?

This question is of some importance in commercial contracts for plant and equipment. It may be that newly installed production plant has to be shut down during the warranty period because of defects. If, for example, the shut-down is of a substantial nature, for example, for two months, the purchaser might argue that the warranty should have been for twelve months of commercial use, and that one fifth of this has been lost. In many engineering contracts one now finds provisions which extend the warranty by the same amount as the period(s) of non-use.

Of what length is the warranty carried by replaced parts?

This crucial question carries two possible answers, which may be illustrated by Figure 6.2. Everything depends upon the precise wording of the warranty. It may be that the period of the warranty is a single period which commences upon a given date, and ends upon a given date, and which applies equally to all of the equipment, including any replaced parts. Or, alternatively, it may be that the warranty which applies to replaced parts is a fresh period of equal length to the original warranty, but starting on the date of replacement of the part in question.

Quality and fitness for purpose 105

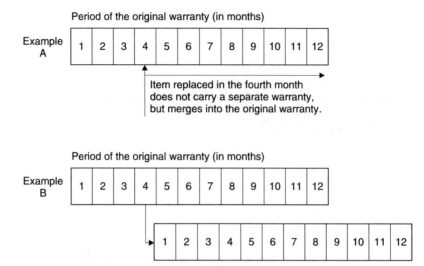

Figure 6.2 *Examples of different types of warranties*

Which party to the contract is entitled to make the choices of remedial action if defects arise during the warranty?

The answer to this question, again, affects the potential value of the warranty, and the risks of the contractor. An item which is under warranty might suffer from a defect which can either be put right by a repair or alternatively by a replacement of the item. Replacement is often more desirable from a purchaser's point of view, but this is not invariably the case, since the purchaser may find that use of the item is disrupted to a lesser extent by an on-site repair. If the purchaser has control of the way in which the warranty is drafted, the purchaser will often take good care to use a phrase such as 'repair or replace at the purchaser's discretion'. A contractor will be aware that the risk of incurring unnecessary costs is lessened if the choice of remedial action rests with the contractor.

7 Liabilities, exclusions and indemnities

Different kinds of liability

In the previous chapter the form of liability known as 'defects liability' was looked at. Defects liability is contractual in nature, by which it is meant that it arises out of a contract existing between two parties. Where a warranty is given by a manufacturer to an end-user who does not buy directly from that manufacturer, the liability is still contractual: the warranty is enforceable for either of two reasons. It may be a transferable warranty. The transfer of it then operates as a form of assignment between the transferor and the transferee, so that the transferee enjoys the same rights as the transferor had. Alternatively, the warranty may be of the kind which is offered to the first user of new goods which have been purchased through a distributor or other outlet. This often takes the form of a card or leaflet enclosed with the goods, and it functions as a form of 'collateral contract' between the manufacturer and the end-user. A collateral contract is a contract existing side-by-side with the main contract, connected with it, but at the same time distinct. If a main contractor warrants work and materials which are included in a project, there may still exist at the same time a collateral contract between the end-user and a manufacturer of components or equipment included in the project. Whether a collateral contract exists or not is a matter of fact and law: the manufacturer must have done something to create in the mind of the end-user the belief that a collateral contract exists, and there must have been some reliance by the end-user upon that belief.

But defects liability is by no means the only possible liability that can arise in an engineering contract, and the conditions of each contract should also take account of other forms of liability. Some of these liabilities may cause death or personal injury to employees or to third parties; there may be damage to the property of either the employer or the contractor, or to the property of third parties. It is important that each engineering contract should not only deal with such liabilities, but should also provide for the taking out and maintenance of the appropriate policies of insurance. If either party to the engineering contract breaks a terms of the contract, and thereby causes loss or damage to the other party, such as the physical destruction of property, the liability may take two possible forms: the first will be the more obvious, which is that of *breach of contract* whereby there will have been a breach of one or more specific undertakings to take due care and skill. But another form of liability may exist side-by-side with the first, and this second form is known as liability in *tort*.

Liability in tort includes several different forms of tort or civil wrong, such as negligence, nuisance, defamation, breach of statutory duty, etc. The different kinds of torts have grown up separately and each have their own rules, but in each case it can be said that the liability in question stems from a failure by one person to take care to avoid injury loss or damage to the other. Unlike the law of contract, in tort there need be no pre-existing legal relationship between the persons concerned. So liability which is between unconnected parties, who are not bound to each other by any contract or collateral agreement, will arise under the law of tort. Even if the parties do have subsisting contractual arrangements, the same incident may give rise to liability both under the law of contract and under the law of tort. This is because it is possible to break a specific contractual duty *and* the duty to take care under the law of tort, at one and the same time.

While these points may seem somewhat academic, they are by no means so. The terms of a contract may, for example, limit liability which arises under the contract, but unless the wording of the contract also deals with tort, the limitation of liability will not apply to a liability arising under the law of tort. The result of this will be that the co-existence of both forms of liability arising out of the same incident will, in effect, bypass the limit of liability.

The same considerations apply to *indemnities*. Indemnities are undertakings to keep another person free from loss, or to repay the losses of that other person. A contractor may, for example, undertake to indemnify the employer against any claims in respect of injury or damage to any third party arising during the course of the works. If a claim is made by a third party, and the injury or damage has arisen solely due to a breach of a contractual duty by the contractor, few problems are likely to arise: the indemnity will be applied, and the loss will fall upon the contractor or the contractor's insurers. But if the injury or damage arises due to *negligence* (which is a form of tort), then complications can arise. In claims for negligence during construction works, it is quite common for *both the employer and the contractor* to be found to be partly to blame for an incident. If a claim is made by a third party in such circumstances, the chances are that the indemnity will be held not to apply. This is because either the wording of the indemnity will specifically exclude application in such cases, or it will be *implied* by the courts that it is excluded in such circumstances.

This principle is illustrated by the case of *Walters v. Whessoe Ltd and Shell Refining Co. Ltd* (1960), in which Whessoe Ltd, the contractor, had made an agreement with Shell Refining Co. Ltd, containing an indemnity clause under which the contractor was to indemnify the employer against *all claims arising out of the operations*. Mr Walters, an employee of Whessoe Ltd, was killed in an accident on site. His claim was brought under the law of negligence, and both Whessoe Ltd and Shell Refining Co. Ltd were held partly to blame. Shell sought to enforce the literal wording of the indemnity clause, but were unable to do so. The court held that as the indemnity clause did not expressly, or by implication, cover the possibility that both employer and contractor might become liable jointly to a third party in tort; the indemnity did not apply in such circumstances.

These considerations, arising out of the distinction between and interaction between claims in contract and in tort, make the subject of indemnity clauses an excruciatingly difficult one, and unless the clauses are very precise and specific, advice should be sought both from a legal point of view and from the point of view of one's insurer at the time of negotiation of the contract terms.

Negligence

Of all the branches of the law of tort, negligence is the one which occurs most frequently in commercial, engineering, or construction contracts. It may arise wherever there is an accident causing personal injury or death, or loss of, or damage to, property. It arises wherever any person is in a position in which care is expected of them, so as not to cause such injury, loss or damage. This is known as the 'duty of care', and has no relationship at all to the law of contract: it exists simply because the possibility of injury, loss or damage can be foreseen by reasonable people. A driver of a vehicle, for example, is under a duty of care at all times; so is an operator of plant or equipment; so is an erector of scaffolding; and so on. In its modern form, the duty of care was given shape in a number of cases concerning consumer goods such as clothing, food and drink, but these are of little interest to the engineer, and need not be rehearsed here. However, some of the more recent and relevant examples follow.

In the case of *Muirhead v. Industrial Tank Specialities Ltd and Others* (1985) the main contract was for the installation of a tank, complete with pumps for the circulation of water. The electric motors for the pumps turned out to be unsuitable for the English voltage range, and the pumps cut-out

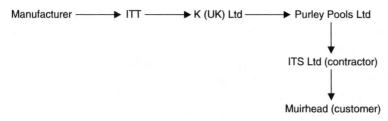

Figure 7.1 *Muirhead v ITS Ltd and Others* (1985)

when there were voltage surges, which fell within the permitted tolerances in England, but could exceed 240 volts by as much as 6 per cent. This cut-out caused damage. The purchaser, naturally, had a valid claim against the main contractor, on grounds of unfitness for the required purpose. However, the main contractor became insolvent, and the judgment which was for a sum in the region of £150 000, was unsatisfied. It is in such circumstances that a purchaser may turn to the law of tort, as an alternative means of recourse. The essential difference, however, is that whereas with a contractual duty one simply looks to the terms of the contract, in tort one must first of all identify the person who has committed the tort, that is the person with the duty of care, and in breach of that duty of care. In this case there was a long chain of manufacturing and distribution before the pumps reached their final destination, as can be seen from Figure 7.1.

In the event, it was held that a claim could be made against the manufacturer of the electric motors for the pumps, on the basis that the manufacturer knew that the motors were to be used in pumps in the United Kingdom, to circulate water in circumstances of the kind applicable here. It followed that the manufacturer had a duty to take care to make sure that the pumps were suitable for use in the United Kingdom.

One of the differences between a claim made under the law of contract and one made under the law of tort is in the method of calculating damages. In both cases the starting principle is the same, which is that the court will try to put the buyer into the same position, by the award of damages, as if the wrong or breach of contract, had not been done. However, damages will only be in respect of losses or damage of a kind which the parties would reasonably have foreseen or had in mind either at the time of the making of the contract, or in the case of negligence, at the time of the negligent act. There are subtle differences in the test for breach of contract and the test for damages in tort, but the most obvious difference is that two parties to a contract will normally have a closer relationship than parties to an action purely in tort, without any contractual relationship, and consequently the loss or damage that might have been expected will tend to be greater in a claim for breach of contract. Leaving aside any possible limits of liability which might exist in the terms of the contract, the contractual relationship will often be such that a contractor will be more likely to foresee the full extent of loss to a customer due to breach of contract. This is why, in principle, damages for breach of contract can include damages for such economic losses such as loss of profits and loss of production. However, when a claim is made in tort, such as a claim for damages for negligence, purely economic loss is often thought not to be reasonably foreseeable between two disconnected parties. Physical damage, of the kind that may well arise if there is an accident caused by negligence, can usually be claimed in the law of negligence. Financial loss, which is immediately associated with any physical damage, can also be claimed: a person who suffers physical injury due to the negligence of another can not only claim for pain, suffering and any physical impairment, but also for medical expenses and loss of earnings. However, if a claim is for purely economic loss, without there being any physical damage, then this will usually be disallowed under the law of negligence.

In the case of *Simaan General Contracting Co.* v. *Pilkington Glass Ltd* (1988), the contractor, Simaan, was the main contractor to a customer in the Middle East. The project required glass of a uniformly green colour. This was purchased by a sub-contractor to Simaan, from Pilkington Glass Ltd. The glass was rejected by the customer of Simaan on the ground that it was not uniformly green, but contained shades of red in some places. The rejection of the glass caused economic loss to the main contractor, which the main contractor sought to recover. An action for breach of contract against the sub-contractor was not possible in this case, so Simaan brought its claim against Pilkington Glass Ltd. However, no contract existed between the two parties, so a claim had to be made in tort, for negligence. The problem for Simaan was that there was no question of physical damage: the product was not damaged, and did not cause any damage; it simply happened not to conform to the standards of the end-user. It was held by the court that the claim of Simaan would be disallowed. Simaan had specified Pilkington Glass, but they had not negotiated directly with them or relied upon any form of collateral warranty or undertaking or other relationship with Pilkington Glass Ltd. It is the element of reliance and foreseeability which gives rise to a liability to pay compensation for the economic loss of another party, in the law of negligence, and this element was absent in this case.

One of the few cases in which a claim for purely economic loss by a party who had no contract with the party being sued was allowed, was the case of *Junior Books Ltd* v. *The Veitchi Co. Ltd* (1982). This case involved the relationships shown in Figure 7.2.

Liabilities, exclusions and indemnities 111

Figure 7.2 *Junior Books Ltd* v *The Veitchi Co Ltd*

The claim concerned an allegation by the employer that the nominated sub-contractor had been negligent in the laying of the floor of a factory. As the employer was not in any contractual relationship with the sub-contractor, the action was brought in tort, and not in contract. The case was a Scottish one, but it was accepted that there were no relevant distinctions to be made between the laws of England and Scotland for the purposes of the case. The issue before the House of Lords was not the facts of the case, but the question of whether, even if negligence could be proved, the damages claimed, which were purely in respect of economic loss, were of a type which could be awarded under the law of negligence. The floor of the factory was defective, but it had not actually injured anybody, and was not in itself a danger to any person or to any other part of the premises. It was simply a product which was not in good condition and was not worth the price.

A claim against the *main* contractor would have raised commonplace issues, and would not have been worth the costs involved in an appeal to the House of Lords. The main contractor would have been liable on the grounds that the floor was not of the required quality and was not fit for the required purpose. However, the claim was made against the *sub-contractor*, and as there was no collateral warranty, the case had to be brought under the laws of tort. That is why the issue of purely economic loss was relevant. In this particular case, exceptionally, the House of Lords held that a claim *could* be made for such loss. The reason for this was, presumably, that a relationship of *nominated* sub-contractor gives the employer a much closer relationship with the sub-contractor than is the case where the sub-contractor is not nominated. In many cases, where the sub-contractor is nominated, there are direct negotiations about such matters as price and specification between the employer and the sub-contractor, and this can create the reliance and the necessary foreseeability of loss, which will entitle the employer to claim in tort for purely economic loss. Having said this, such claims are seldom likely, in the present state of the law, to be successful.

Product liability

Product Liability is a relatively new form of liability which is similar to liability in tort, in that it can, and usually does, involve third parties. However, its origins are different from the origins of tort, and indeed stem partly from the inadequacy of the laws of tort to deal with certain types of claims. Not all claims for product liability will have anything at all to do with engineering contracts, but they may do so, and such claims may, hypothetically, arise as follows. A company which produces consumer goods such as motor cars may purchase services or equipment for the engineering of its products. There may be a fault or defect in those services or the equipment purchased, and this may result in defects relating to the safety of the end product. An unsafe end-product which causes injury to persons or damage to private property can give rise to a claim for product liability, and this may be made by any person suffering the injury or damage. Such a claim would have financial consequences for the company which produced the unsafe consumer items, and the financial consequences almost certainly would not stop at the compensation paid out on the claim itself: the costs of *recall* of the potentially unsafe items and possibly the costs of re-design of the entire product, could be far greater.

Definitions and origins of Product Liability

Product Liability can be defined as liability for injury or damage caused by defective or badly designed products, arising irrespective of the existence of any contract between the parties, and irrespective of any need to prove negligence on the part of the party against whom the claim is made.

It is the point about the absence of any requirement of negligence that distinguishes Product Liability from the tort of negligence. This distinction is deliberate: Product Liability originated in the United States, and came into existence in member states of the European Community due to a European Community Directive passed in 1985. The reason in both cases, whether in the United States or in the European Community, was that it was felt that the burden upon a claimant of having to prove negligence was, in many cases, too great to provide fair protection to the public against manufacturers of unsafe goods. In order to win a claim for negligence, as the *Muirhead* case (discussed earlier in this chapter) shows, one must first of all pin-point the exact cause of the injury or damage, and then identify with great precision the party responsible for this. One must then show that the injury or damage resulted from a breach of a duty of care owed by one party to the other. The interaction of these points can create great difficulties for claimants, particularly where injury or damage is caused by multi-component items – such as motor cars or aircraft, or by medical or pharmaceutical items, where there are many factors, such as dosage, or recommendation by a particular practitioner to a particular patient – which can complicate the issue.

In the United Kingdom, the European Community Directive on Product Liability was implemented through the Consumer Protection Act 1987, which came into force on 1 March 1988. Under this Act, virtually any physical item is a 'product', the exception being buildings (although building materials are products). The Act states, in section 2.1: '... where any damage is caused wholly or partly by a defect in a product, every person to whom subsection (2) below applies shall be liable for the damage.' Section 2.2 states that the following persons are liable:

(a) the producer of the product;
(b) any person who, by putting his name on the product or using a trade mark or other distinguishing mark in relation to the product, has held himself out to be the producer of the product;
(c) any person who has imported the product into a member State from a place outside the member States in order, in the course of any business of his, to supply it to another.

Analysis

This adds an extra layer of potential liability to the laws of contract and tort. It does not alter those laws, but it does give to consumers additional protection. Only consumers, and not businesses, may bring claims under the laws of Product Liability. A business which has to pay compensation under such a claim, or which has to withdraw or redesign a product, cannot claim under the law of Product Liability, but can use its contract, or the law of negligence, against a supplier to it of any goods or services which can be shown to have caused the fault in the product.

Potentially any area of industry can be affected by the new laws of Product Liability, and all engineering contracts should anticipate any risks and problems which might arise. As Figure 7.3 of this text shows, materials and components can give rise to Product Liability just as much as end-products, and terms of engineering contracts should allocate responsibilities as to design, interface of components, choice of materials, testing and safety precautions, and insurance against liability, accordingly.

Two further aspects of the Product Liability legislation must be looked at here, since they have engineering implications. One of these is the definition of the word 'defect' in section 3 of the Act:

114 Liabilities, exclusions and indemnities

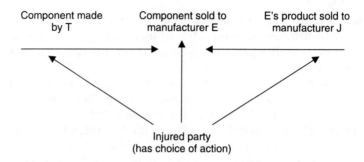

Note: Under the Consumer Protection Act 1987, the injured party must be an individual suffering injury or damage to private property. He may claim against any producer, importer or 'own brander' of the product. The product includes all components and materials in it. The company which buys a component, and which is sued for damages by an individual, may use its contract with the component manufacturer to 'pass back' liability. Failing this, it could use the laws of negligence.

Figure 7.3 *Parties to product liability*

Subject to the following provisions of this section, there is a defect in a product for the purposes of this Part if the safety of the product is not such as persons generally are entitled to expect: and for those purposes 'safety' shall include safety in the context of risks of damage to property, as well as in the context of risks of death or personal injury.

Section 3(2) goes on to explain that safety takes into account:

(a) the manner in which a product is marketed, and any instructions that are given with it;
(b) the things which might reasonably be expected to be done with the product;
(c) the time when the producer supplied the product to another.

This definition and the qualifications attached to it make it clear that there could be Product Liability in respect of a product which is correctly made, but which lacks information or instructions which would enable it to be safely used.

The other aspect of the Product Liability legislation which all producers of goods should study carefully is the statutory defences which are provided in the Act. They have been included because the nature of Product Liability is *liability without fault*. For example, each company which brands an item is responsible for that item and every material and component in it. This means that in many cases companies will be liable in respect of items which they did not make an did not design. However, common sense requires that the liability must end somewhere, or have reasonable, practical exceptions. For this reason the following defences exist. These are that:

(a) the defect is attributable to compliance with any requirement imposed by or under any enactment or with any Community obligation;
(b) the person proceeded against did not at any time supply the product to another;
(c) the supplier was not in business;
(d) the defect did not exist in the product at the relevant time;
(e) the state of scientific and technical knowledge at the relevant time was not such that a producer of products of the same description as the product in question might be expected to have discovered the defect if it had existed in his products while they were under his control;
(f) the defect
  (i) constituted a defect in a product ('the subsequent product') in which the product in question had been comprised, and,
  (ii) was wholly attributable to the design of the subsequent product or to compliance by the producer of the product in question with instructions given by the producer of the subsequent product.

Most of these defences do not require explanation, but those numbered (e) and (f) need to be carefully considered, with examples of their possible application. Defence (e) is sometimes called the 'development risk' defence, or 'state of the art' defence. It could apply to a drug, or a chemical, for example, which after many years of research proves to have harmful side effects. If a producer is permitted to place it on the market before any other producer of similar products, or the actual producer itself, has any knowledge of possible defects, a defence will exist. It is not enough for a producer to claim to be unaware of any likely defects. The producer must show also that no other producer was or could reasonably have been aware of the harmful side effects.

Defence (f) is a defence for certain component makers, or makers of interface items. A maker of brake linings, for example, may make exactly what has been specified, but the linings may be unsuitable for the application or system into which they are to be fitted. If the entire system proves to be defective, the defence for the maker of the brake linings will be that they did not design the system (called the 'subsequent product'), and that the defect was not caused by the producer of the linings, but by the producer of the entire system. This defence is sensible, and in accordance with the principles of design responsibility which have been already discussed in an earlier chapter of this book. What it does emphasize is the importance of clarity as to who is responsible for the design of an item.

Limits of liability

Having discussed the different types of liability which may exist in an engineering contract, it now becomes clear that one of the functions of a commercial contract is to place limits upon the liability which may come into being. The question is whether or not this is legally possible, and if so, to what extent. There is also the commercial question of whether or not limits of liability are desirable, and although this is very much a value-judgment, it merits some discussion here.

A purchaser of goods or services or works might, as an initial reaction, argue that limits upon the liability of a contractor are completely undesirable, and in no circumstances to be permitted. However, what such a purchaser will have to confront sooner or later is the possibility that unlimited liability may have an adverse effect upon prices, or may even lead to reluctance of certain kinds of contractors to undertake certain kinds of work of a commercially risky nature. It is with these thoughts in mind, no doubt, that several of the major forms of engineering contract published by the institutions contain quite far-reaching limits upon the liability of the contractor. A purchaser may, for example, require engineering works to exploit a new market with a new 'niche' product. The perception of the market, and the assessment of the possible requirements, and the decisions as to how to approach the project and to cater for risks that may arise, are mainly the purchaser's judgments. The rewards arising from a successful project belong mainly to the purchaser. Looked at in this light, it is not difficult to appreciate why the contractor, although accepting some liability for delays and defects, will not wish to accept unlimited liability. The purchaser, in such a case, is arguably in a far better position to take precautions against the risks, and the costs to the purchaser of doing so may be proportionately less than they would be to the contractor.

What techniques are there of limiting liability?

These are more numerous than is often realized. The ways in which liability may, theoretically, be limited, by agreement between the parties, are by reference to: *time limits*; *monetary limits*; and *limits upon types of liability*. Within these categories there may be further sub-categories; for example, *monetary limits* may be agreed either as sums expressed in a particular currency, or as a percentage of the purchase price or contract value. The *Limits upon types of liability* that will be imposed on, or accepted by, either or both of the parties very a great deal from contract to contract, and have a relationship with the variable characteristic of a 'warranty', which were discussed earlier. It is quite common for *limits upon type of liability* to be expressed as a limit upon or exclusion of 'consequential' loss or damage, and the frequent use of this expression raises questions as to its precise meaning. To appreciate this fully, one must enter a linguistic and conceptual maze. It may be helpful first of all to inquire what the parties actually intend when drafting such a provision.

'Consequential' loss or damage

When a clause excluding or limiting liability for 'consequential' loss or damage is drafted into a contract, either or both of the parties, but more commonly the contractor, will be attempting to exclude liability for *economic loss*. The argument is similar to the one already discussed in relation to damages in the law of tort. The difference, however, is that whereas economic loss is seldom the subject of an award of damages in tort, it is perfectly normal under the law of damages for breach of contract. That is why parties to a contract may draft a clause into the contract limiting or excluding such liability.

The problem, such as it is, has arisen because of a linguistic misunderstanding: many people in commerce, who are not lawyers, believe that 'consequential loss' is the same thing as 'economic loss'. So, for example, they believe that to exclude 'consequential loss' will have the effect of excluding liability for loss of profit. The reality is that if the parties wish to exclude liability under the contract for loss of profit, they must actually say it in so many words. If they prefer to use the expression 'consequential loss', then they should define it to describe precisely what they intend it to mean. In the absence of definition, a court is bound by precedent to take 'consequential loss' to mean *only the more remote damages arising from a breach of contract*; the damages that the parties would not normally be aware of, unless they had received advance warning that they were likely to arise. Loss of profit (as well as other common financial losses such as 'waiting time', 'idle time', and money paid to third parties due to delay) is not normally classified as 'consequential loss', because it is commonplace, and not at all remote, except in cases of special profit.

In the case of *Croudace Construction Ltd* v. *Cawoods Concrete Products Ltd* (1978) the defendants thought that they were protected from liability for delay and defects by a clause which excluded liability for 'any consequential loss or damage caused or arising by reason of late supply or any fault, failure or defect in any material or goods supplied by us'. The damages claimed were in respect of money necessarily paid to the workforce which had been kept idle due to late delivery of masonry blocks. The Court of Appeal held that the damages claimed were not so remote as to come under the legal meaning of consequential loss or damages, and therefore liability was not excluded.

To what extent can liability be limited or excluded?

At one time there was only one rule that a party to a contract had to satisfy in order to be able to rely upon an exclusion or limitation clause, the rule of construction (an example of which was illustrated in the previous paragraph). Judges would read such a clause strictly, and would give it no greater meaning than it actually conveyed. This rule, or 'hurdle', which the party seeking to rely upon the clause must surmount, still applies and can be seen in its application to consequential loss, as well as in the fact that clauses limiting liability in contracts do not necessarily limit liability in tort, unless they make it clear that that is the intended meaning.

Apart from this hurdle, there are two more hurdles which, in the modern state of the law, must also be considered; the UNFAIR CONTRACT TERMS ACT 1977, and the EC Directive on Unfair Terms in Consumer Contracts 1993. Although there is some possible overlap between these laws, their scope and application also contains important differences. This is illustrated by Figure 7.4

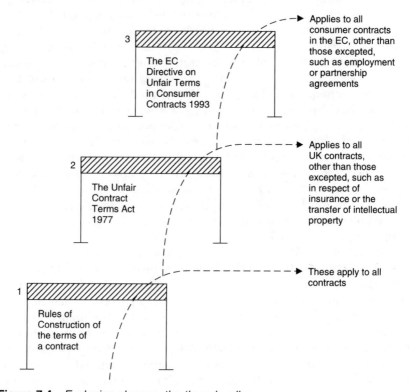

Figure 7.4 *Exclusion clauses: the three hurdles*

As can be seen, rules of construction of terms of contracts apply to all contracts, without exception. They are simply part of the power of the courts to interpret actual bargain reached between two parties. Courts have a tendency to use rules of construction to confine exclusions and limits of liability fairly narrowly. This is understandable, since the effect of such clauses is to take from a party the rights that that party would enjoy under normal rules of contract and tort. The Unfair Contract Terms Act 1977 was passed because it was felt that the powers of the courts did not go far enough in controlling possible abuses. It was felt that some form of statutory limits upon the ability of parties to draft into contracts exclusion clauses were also much to be desired. This Act is, unfortunately, an excessively complex piece of legislation, and it is not possible to examine all its implications in this work, even if those implications were fully known and understood. However, the following summary may be found useful. The scheme of the Act is such that some types of exclusion or limitation of liability are of no effect at all. The exclusion clause is, in effect, void, and the contract is read without it. Other types of exclusion or limitation of liability may be tolerated by the courts, but this will depend upon them satisfying the test of reasonableness.

Some clauses or terms attempting to exclude or limit liability are ineffective

These are those which
- (a) attempt to limit or exclude liability for death or personal injury caused by negligence;
- or (b) attempt to limit or exclude the statutory rights of a person dealing as a consumer;
- or (c) attempt to exclude the implied undertakings as to title given by a seller or supplier of goods.

Under the Unfair Contract Terms Act 1977, *some* terms attempting to limit or exclude liability are not entirely ineffective, but are only effective **to the extent that the term in question is reasonable**

These are those which
- (a) attempt to limit or exclude liability for negligence *not* causing death or personal injury: for example, damage to property;
- or (b) attempt to limit or exclude the statutory rights of businesses;
- or (c) attempt to use standard conditions to exclude or limit liability generally, or to claim to be entitled to perform insufficiently or not at all.

In the United Kingdom, as a result of the passing of the Unfair Contract Terms Act 1977, contracts must not only clear the usual hurdle of the interpretation and construction of their terms, but they must also clear the additional hurdle of having any exclusion or limitation of liability clauses tested for validity or reasonableness by the courts. So, for example, in the case of *St Albans City and District Council v. International Computers Ltd* (1994), where ICL had supplied software to a local authority to deal with the calculations needed for making the community charge (the 'poll tax'), there was a term in the contract limiting the liability of ICL to the sum of £100 000. In the event, there was an error in the software which caused to the local authority a loss of around £1 300 000. The High Court held that the limit of liability did not satisfy the requirement of reasonableness under the Unfair Contract Terms Act 1977. It was held that the contract was made on the standard terms of ICL, and that any limits of liability in such terms have to be reasonable. The grounds for the decision were the fact that ICL was considered to have had greater bargaining power than the local authority, and ICL was in a better position to carry the loss through its insurance.

The EC Directive on Unfair Terms in Consumer Contracts 1993

If a contract is one between a consumer and a business, and if it does not fall into one of the exempt categories, then it must clear yet another hurdle. This is the EC Directive on Unfair Terms in Consumer Contracts 1993, which was implemented in the UK by regulations in 1995. The chances of an engineering contract of the kind under discussion in this book falling within the EC Directive and Regulations are comparatively slim, because the overwhelming majority of such contracts are made with businesses as the purchaser or employer, and do not involve a consumer. Nonetheless, the EC Directive requires a brief mention in this work, if only to put it into perspective – and because certain types of engineering contracts, (for the installation of heating or for the installation of telephone and computer systems, for example) may well be made with private consumers.

The difference between the EC Directive and the Unfair Contract Terms Act 1977, is that: (a) the earlier Act is United Kingdom law, whereas the Directive is European Community law; (b) the earlier Act applies to business contracts as well as to consumer contracts, whereas the EC Directive only applies to consumer contracts; and (c) the earlier Act does *not* control all clauses of a contract, but only those clauses which are about limits or exclusions of liability, whereas the EC Directive subjects all terms of consumer contracts to the same tests of fairness. The tests of fairness under the EC Directive are summarized in article 3(1):

A contractual term which has not been individually negotiated shall be regarded as unfair if, contrary to the requirement of good faith, it causes a significant imbalance in the parties' rights and obligations arising under the contract, to the detriment of the consumer.

At the time of writing, the regulations implementing the EC Directive have only been in force for a very short while, and therefore there have been no test cases, and consequently no judicial indications of how the tests of the Directive will be applied. What can safely be said is that a completely new dimension has been introduced into those contracts which are made between businesses and consumers. The old question of whether or not an exclusion clause is reasonable is one with which lawyers and the courts have many years of experience in dealing; the new question of whether or not terms of a consumer contract meet the requirements of good faith, or cause a 'significant imbalance' in the parties' rights and obligations, raises issues beyond our present experience, and only time will tell how they are to be resolved.

Some legal questions answered

Is there any advantage in bringing an action for damages under the law of contract rather than tort, or vice versa?

This is a question which is often confronted by the lawyer representing a claimant. It can only be decided after consideration of a number of relevant factors. Under a contract between the parties, liability is usually easier to establish, since, for example, if a machine is unfit for the agreed purpose, the seller is liable for damages for breach of contract without the need for the buyer to prove negligence on the part of the seller. Claims in tort are therefore normally only made if there is some good reason why a claim in contract will not succeed.

What possible reasons are there why a claim made under a contract might not succeed, although a claim in tort might be successful?

Some of the cases looked at in the section of this chapter dealing with *negligence* help to answer this question. In the case of *Muirhead* v. *Industrial Tank Specialities Ltd and Others*, the reason why, after having sued the main contractor for breach of contract, Muirhead pursued his claim against other parties, in tort, was that the main contractor had become insolvent. There was no particular advantage in claiming in tort, and in fact the damages awarded in tort were far lower than those that had originally been awarded for breach of contract, since the laws of tort excluded damages for that part of Muirhead's loss which could be called purely 'economic loss'. However, the action against the other parties could only be brought in tort, because no contract existed between Muirhead and the manufacturers in question.

The case of *Junior Books* v. *The Veitchi Co.* Ltd (1982) raised intruiguing questions as to why Junior Books should have taken the difficult route, of suing a sub-contractor in tort, when the easier route of suing the main contractor was presumably available. No answers to this question were given in the judgments of the courts, but speculation has continued ever since. One possibility is that the terms of the contract might have contained limits which would have been effective to limit liability in contract, but not effective to limit the liability of sub-contractors in tort. Another possibility is that purely commercial considerations may have meant that Junior Books did not wish to sue its main contractor. A third possibility which has been canvassed is that Junior Books may have started a claim against the main contractor, and may have reached an out of court settlement of this claim. They may then have realized that the defects were far more serious and far more expensive to put right than had originally been imagined. However, if properly drafted, the out of court settlement would have been final and binding between the parties. It need not, however, have involved the sub-contractor, and this would have left it open for Junior Books to pursue a claim against a sub-contractor, in tort, in the hope of being able to recover the full extent of its damages.

What these cases tend to show is that although parties who are not connected by contract will be unable to make a claim against each other except under the laws of tort, those parties between whom there is a contract do not normally choose the tort route unless, for some reason, the contractual route is not available. Quite apart from the burden of proving negligence, and quite apart from the fact that damages for economic loss are difficult to obtain in tort, there is also the possibility that a defendant in an action brought in tort will raise the defence of *contributory negligence*.

What is contributory negligence?

Contributory negligence serves as a defence or partial defence in an action brought under the law of negligence. The defence is that the injury, loss or damage was caused to a greater or lesser extent, not only by the negligence of the defendant, but also by negligence on the part of the person making the claim. Contributory negligence can reduce the damages payable to the claimant to whatever extent the court thinks that the claimant has contributed to his or her injury or loss.

Can companies bring claims in respect of Product Liability?

Not as such. Only individuals suffering injury or damage to private property may bring such claims. However, the laws of contract and of tort will allow a company against which an individual has made a claim for Product Liability to recoup the loss against another company which can be shown to have caused the loss. If Company A manufactures 'widgets' for Company B, to go into consumer products manufactured and branded by that company, and if the widgets prove to be defective and cause harm to a consumer, the consumer is most likely to sue the Company B, whose brand name appears upon the product, and may obtain either damages or an out of court settlement from the brander of the product. Company B may then use its contract with Company A to attempt to recoup its loss. The same principle would apply to any loss caused by having to withdraw a product in respect of which a safety problem had been identified.

Does Product Liability only apply to consumer goods?

No it does not. Claims are restricted to consumers, but the types of goods which may give rise to a claim are not at all restricted. In practice, because statistically most consumer claims tend to be most commonly involved with consumer goods, such goods are the most likely to be the subject of a claim. However, in theory, any goods, whether consumer goods, or aircraft or factory tools and equipment, or heavy duty plant and equipment, could become the subject of a Product Liability claim. The main reason why factory equipment is less likely to be the subject of a claim for Product Liability is probably because an injured employee will not, in practice, need it: he or she will already have an avenue for a claim under statute, such as under the Employer's Liability (Defective Equipment) Act 1969.

What kinds of exclusions of liability have the courts held to be unreasonable?

One recent example of this has been illustrated previously in the case of *St Albans City and District Council* v. *International Computers Ltd*. This case is by no means the first to hold that a monetary limit of liability is capable of being unfair. The limit may be unfair if expressed as too low a sum, compared with the actual and foreseeable loss resulting from a breach of contract. It may also be held unfair if expressed as the value of the contract goods (or a fraction of their value). This was the case in *Mitchell (George) (Chesterhall) Ltd* v. *Finney Lock Seeds Ltd* (1983) in which liability for defective seeds was restricted to the price of the goods sold (£200), but the House of Lords allowed a claim by a commercial farmer for damages in the region of £60 000. However, monetary limits have not been the only ones held to be unfair. In *Rees Hough Ltd* v. *Redland Reinforced Plastics Ltd* (1984) a contract term, which in effect restricted the warranty in respect of defects to a period of three months and which excluded all other liability, was held to be an unfair term.

Perhaps the most singular example so far in an engineering contract is the case of *Stewart Gill Ltd* v. *Horatio Meyer Ltd* (1992), the facts of which have already been given in Chapter 3 of this book. What makes this case special is the fact that the offending clause did not at first sight appear to be an exclusion or a limit of liability; it did not apply monetary limits to liability or time limits; nor did it restrict or limit the types of liability in respect of which claims could be made by the purchaser. What the clause did was to state that the customer could not enforce any of his remedies by withholding money under a set-off or counterclaim. However, the Unfair Contract Terms Act 1977 was carefully drafted so as to ensure that what could not be achieved by direct methods would also be difficult to achieve by indirect means. If it is capable of being unfair to deprive a purchaser of legal rights under a contract, it is also capable, under section 13 of the Act, of being an unfair term if the term makes the right of the purchaser subject to restrictive or onerous conditions. To state in contract conditions that the purchaser may have a right in respect of defective goods, but will not be allowed to enforce that right by way of set-off against the price is tantamount to saying 'pay first and argue afterwards'. This case is an example of how far the law has progressed towards controlling unfair terms, even in contracts made between businesses.

8 Ownership of goods and intellectual property rights

If you wish to build a better bridge, start from first principles. Do not look at any bridge that has been built before
　　　　　　　　　　　　Attributed to Isambard Kingdom Brunel

Ownership of goods and materials

The ownership of goods and materials is of considerable significance in an engineering contract. The contract will almost invariably contain one or more clauses dealing with this matter. This part of this chapter will examine why ownership matters, and consider the different ways in which the parties to such a contract may express terms about ownership. As with many other terms of a contract, it is a matter for negotiation and for the parties to strike a balance of commercial convenience.

Most engineering contracts other than those which are purely for services, such as maintenance or for research and development, involve the transfer of materials. The value of these materials may vary, from constituting a major factor in the make-up of the contract, to having only a marginal value when compared to the design or labour element of the contract. In all cases, however, it can be said that there is a tangible *asset value* to be taken into account when drafting and negotiating the terms of the contract.

Ownership of goods and materials helps one or other of the parties to secure this asset value, particularly when combined with insurance of the goods. It cannot provide a complete security, because there are a number of ways in which an owner of goods can be quite legally deprived of the ownership of goods during the course of commercial transactions. This is because, in the ordinary course of business, the law tends to favour bona fide third party sub-purchasers against the original owners of goods. Thus, a manufacturer of goods may deliver them to a main contractor, and may state that they remain the goods of the manufacturer until they have been paid for. If, however, the main contractor sells the goods on to his own customer, or builds the goods into the property of his own customer, the statement as to the ownership of the manufacturer will have no effect, and the ownership of the goods will be lost to the manufacturer, in favour of the customer. This is one of the major weaknesses in the notion of retention of ownership as a security for payment. Having said this, it remains the case that, by and large, it is more valuable to have ownership of goods than not to have it, particularly in those instances where there is no likelihood of sub-sales to third parties.

Express provisions in contracts about ownership of goods

In general, the law does not contain any restriction upon the right of parties to a commercial contract to state when the ownership of goods is to be transferred. Clauses are drafted with varying provisions and in varying styles. In some contracts the word 'ownership' is used; in others, the expression 'the property in goods', or a similar expression, may be used. The precise choice of words does not appear to matter as much as the need for clarity as to the choice of timing that is being made. Ownership of goods may pass either *before delivery*, or *immediately upon delivery*, or *after delivery*. In each case, the precise event that causes the transfer of ownership must be clearly defined. There is, of course, a link between these terms of a contract and the terms about delivery which were discussed in an earlier chapter. Certain terms about delivery, it will be recalled, imply the transfer of the risk in and ownership of goods at a particular moment. It is important that, if there are two different clauses, the terms about delivery and the terms about the transfer of the ownership or property in goods should be entirely consistent with one another.

Ownership to pass on delivery

This is the standard position in a contract for the sale of goods, and would, in the absence of any indication to the contrary, be implied by law. 'Delivery' means placing the goods at the disposal of the purchaser, so delivery FOB will have the effect of passing property in goods to the purchaser, but delivery to a ship on CIF terms will not, by itself, have this effect, since the goods will not be at the disposal of the purchaser until the required documents have also been handed over. Another interesting exception is in contracts for building or construction work on the site of the purchaser: in such contracts the contractor (or sub-contractor, as the case may be) will often be on site, in a particular area set aside for the use of that contractor, and will often receive goods and materials on site for his own use. In such circumstances, the delivery of goods and materials to site will not, in the absence of terms to the contrary, be sufficient to pass ownership to the site owner or purchaser. The contractor (or sub-contractor) can be said, in such cases, to be delivering *to himself*, rather than placing the goods at the disposal of the purchaser. This principle has often had important consequences, as for example in the case of *Dawber Williamson Roofing Ltd* v. *Humberside County Council* (1979), in which a main contractor became insolvent, and both the site owner and a tiling sub-contractor laid claim at the same time to the tiles which had been brought on to the site. The site owner based its claim upon the fact that it had already made a stage payment to the main contractor in respect of the tiles. The sub-contractor based its claim on the fact that the delivery of the tiles to site was only a delivery of the tiles for the purposes of the sub-contractor, and not a sale to the main contractor. The sub-contractor won the case. As a consequence of this case, parties to such contracts have had to take considerable care over the arrangements as to transfer of ownership, and can no longer take it for granted that goods lying on site have actually been sold either to the site owner or to a main contractor. It should be noted that a clause in the site owner's conditions of contract stating that ownership of materials is to pass on delivery to site or upon payment, whichever is the earlier, will be of use to the site owner, but will not have full effect unless a similar clause also appears in the conditions of contract agreed between the contractor and the sub-contractor.

Ownership to pass before delivery

This is a less usual, and non-standard provision which may be put into an engineering contract, or contract for the sale of goods. At first it may be asked why parties, in particular a seller, should ever wish to contract on such a term. However, where valuable items are being manufactured for a buyer, and where the buyer is funding the manufacture (and possibly the design and research and development as well) by means of advance payments or stage payments, it becomes evident that it is in the interests of the purchaser to secure the *asset value* in the project. A suitable clause which provides that where any payment in respect of goods or materials intended for the contract has been made, the ownership of those goods and materials shall immediately pass to the purchaser, will help to give to the purchaser an element of security. In government defence contracts, the vesting clause sometimes goes even further than this, and transfers ownership of goods and materials allocated to or intended for the project to the purchaser at the moment of appropriation, *whether paid for or not*. Such a clause would, of course, be completely contrary to the interests of the seller or manufacturer in most instances, since the giving up of ownership would not be balanced, from the seller's point of view, by the security of having received payment or part payment. For this reason, such a clause is not likely to be encountered in contracts other than government contracts, where there is little risk of the purchaser becoming insolvent.

If purchasers do decide that the system of payment requires a condition which transfers ownership of goods or materials to the purchaser prior to delivery, the purchaser should not only take great care as to how such a clause is drafted, but should also take some care over the project management aspects of this, such as visits and inspections to ensure that the goods actually exist and that they have been properly identified and labelled, or marked as the property of the purchaser. No matter what the conditions of contract may say, property in goods or materials cannot pass to the purchaser prior to delivery unless there has been some kind of ascertainment of which goods are intended for that particular purchaser. This ascertainment can be either by appropriate identification or marking, or by some form of segregation. Bulk goods intended for different purchasers are a separate issue, now dealt with by the Sale of Goods (Amendment) Act 1995, and will not be looked at in detail in this work.

Ownership to pass only upon payment in full

This third type of condition as to transfer of ownersip in goods and materials is usually intended to counter either one of the types of conditions previously discussed. In many cases, a manufacturer or seller of goods, or contractor, is in a reasonably strong bargaining position, and may be giving some form of credit facility to the purchaser. If this is the case, the seller may not wish the normal rule that ownership passes upon delivery to the purchaser to apply. If such a rule were to apply, then the seller might feel financially less well protected than might be necessary, since the purchaser would now have the asset value upon delivery, but would still not have made payment in return for that value. The most serious danger, from the point of view of the seller, is that a purchaser might become insolvent, or in some other way have difficulty in paying its debts. If this occurs, and if title to the goods has already passed to the purchaser by virtue of delivery, then the seller will be in the position of an *unsecured creditor*. In the majority of cases of insolvency, this in practice means only a small chance of the purchaser receiving any payment. On the other hand, if title to the goods remains with the seller, because of the conditions of the contract, then the seller can at least recover the goods, in the event of being unable to obtain payment. If we take as a possible example the flow chart of events illustrated in Figure 8.1, we can see that in those cases where payment occurs some time after delivery or shipment to the purchaser, the position of an unpaid seller is not entirely

Ownership of goods and intellectual property rights 129

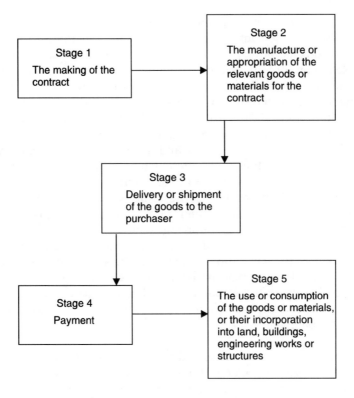

Note: This illustration is only an example of possible stages of an engineering contract. Many different sequences of events are possible, which may affect the arrangements as to payment and transfer of ownership.

Figure 8.1 *Stages of a contract: the passing of property*

secure. If letters of credit are used as a means of payment, then this will alter the position and will provide considerably more security for the seller. However, in 'home' contracts, where letters of credit are not commonly used, the seller may not be prepared to transfer ownership to the purchaser as early as Stage 2, or Stage 3 in Figure 8.1, and may wish to postpone the passing of property until Stage 4: the making of payment.

Retention of title

The postponement of the passing of property in goods to a date or event beyond that which would normally apply under common law is known as the 'Retention of Title'. It can only be done by a most specific clause which is part of the contract and agreed upon by both parties. Many a seller of goods has attempted to create a security for payment by means of a Retention of Title clause contained in an invoice or delivery document, only to find that this is quite useless to retain title, since the document will be a post-contractual document and not part of the contract itself. Retention of Title became extremely popular a decade or two ago, and was widely used in sales and engineering contracts. Of late, its popularity has been moderated, and this may be put down to the fact that at the time of writing, more often than not, conditions of contract are drafted by purchasers rather than sellers. Purchasers have little interest in, and little to gain from, such clauses. Another undoubted factor which has tempered peoples' enthusiasm for Retention of Title clauses has been the practical difficulties of enforcing such clauses. Even in bilateral contracts the difficulties are real enough; in chains of contracts involving sub-contractors and sub-suppliers, the difficulties multiply, since the risks are greater and the contracts are not always consistent with one another. The following check-list of points should always be borne in mind by sellers who intend to place reliance upon Retention of Title to goods as a security for payment:

1 You must take steps to ensure that the contract documentation will stand up to scrutiny. One of the first lines of defence of a receiver or liquidator, if approached for the return of goods, is to argue that there is no evidence that the contract was made on terms incorporating a Retention of Title clause. It is surprising how often this defence succeeds.
2 You will need to be able to identify the goods as being the exact goods sold by you to the purchaser. If goods are generic, such as sand or bricks, or scaffolding, and not easily capable of being identified, this will place serious obstacles in the path of recovery. You can only recover what you can prove to be yours.
3 You will lose title if the goods are validly sold on to a third party in the ordinary course of business. That is the reason why the risks are greater in chains of supply. Your only chance of recovery in such circumstances is if *your purchaser* has also retained title in his contract with *his purchaser.*
4 You will in, any case, lose title (whatever your conditions of contract may say) if the goods or materials are altered after delivery to your purchaser, or manufactured into new products, or if they become fixtures in or part of the fabric of a building or structure upon land. This is one of the rare examples of a rule of common law which is so strongly established that not even the express terms of a contract, apparently, can defeat it.

In summary, this is by no means an easy area of the law or of commercial practice. One judge aptly described it as 'a legal minefield': a warning to all concerned to tread carefully. Further points about this area and about the rules for the passing of property in goods will be looked at in the 'legal questions answered' section at the end of this chapter.

Intellectual property rights

Intellectual property rights are an important aspect of some types of engineering contract or contracts for the sale of goods. Just as with physical property, so with intellectual property we are looking at the *asset value* involved in a commercial transaction. Intellectual property involves different types of assets, compared with tangible physical property, but in many cases this can be of greater value to the owner than the value of any physical property, and for this reason the terms of the contract will be concerned with the creation of, and protection of, such rights. The relationship between the different types of property

Table 8.1 *Different types of property recognized by law*

Real property	Personal property	⟶ Intellectual property
1 Land	1 Goods	1 Rights not created by contract: – patent – copyright – design rights – registered designs – trade marks – service marks – passing off actions
2 Rights in or over land	2 Other forms of personal property: – shares – money – debts – insurance policies	2 Rights capable of being created by contract: – obligation of confidence – agreement not to compete

Note: More than one category of property can exist at the same time: if you produce a drawing or photograph, you may own it as a tangible thing, that is, as *goods*. You may also own the expression of the idea in it, that is, the copyright, which is a form of *Intellectual property*.

with which we are concerned in this Chapter is illustrated by Table 8.1.

In the first part of this chapter, we were concerned with the ownership of that form of 'personal property' which may be described as 'goods'. The property is called 'personal property' to distinguish it from 'real property', or land, and not because it is personal in any ordinary sense of the word. In the second part of this chapter, we are concerned only with the types of property listed in the third column of Table 8.1, that is to say, the different types of intellectual property. They raise different types of issues, because some forms of intellectual property rights are created by contract and some are not.

Intellectual property rights capable of being created by contract

The relevance of these to engineering contracts is clear; secrecy agreements, obligations of confidence, agreements not to compete, and similar agreements can form part and parcel of a contract. Indeed, some form of agreement to keep the terms and details of an engineering contract confidential is standard practice. From a purchaser's point of view, it helps to keep developments secret, so that competitors in the same market do not get to know about them. This is of particular importance if a product being developed is not capable of being patented. From a contractor's point of view, it may be important to keep details of a particular contract, such as price, confidential so that the bargaining power of the contractor with different purchasers is not reduced. A contractor will also have in mind the possibility that tender documents which he provides to the purchaser could, unless circulation is restricted, be sent to other contractors who could use the information to compete more easily against the first contractor, who would already have carried a proportion of the overheads of the preparation for the project. The law of *copyright* goes some way to protect the contractor from this: if the contractor makes it clear that the documents are his copyright, then this will prevent others from copying them. It will not, however, prevent disclosure without copying, so it would, in theory, be possible for the single copyright document to be forwarded by the purchaser to a competing contractor, which could rely upon information in it to create a tender which is different enough not to infringe copyright, but which nevertheless borrows some of the ideas and know-how of the first contractor. Only a secrecy agreement, combined with copyright, would be sufficient to prevent this from being possible. As tender documents are often the ones which parties most wish to keep confidential, it is sometimes necessary to have a secrecy agreement in force before the engineering contract comes into existence.

Intellectual property rights not created by contract

These include patents, copyright, design rights, registered designs, trade marks, service marks, and other intellectual property protected by law, such as the right to prevent another party from passing off his goods as if they were yours. At first it may be wondered why an engineering contract should be concerned with these at all. Patents are not created by terms of any contract, they are granted by the Patent Office, after a successful application by the person entitled to apply for a patent. Copyright is not created by contract; it belongs to the author of an original work, including documents, film, photographs, tapes, etc. So we must ask what any clauses or provisions of an engineering contract which mention such forms of intellectual property are intended to achieve. A glance at Table 8.2 may help to put this into perspective. The horizontal headings represent the different types of commercial contract which are common in everyday commercial dealing. The vertical headings represent different types of contractual clause or undertaking. As may be seen, some of these are common in certain types of contract, but not in others. Secrecy or confidentiality agreements are common in all types of contracts. In engineering and sale of goods contracts some types of clause are fairly common, others less common.

The most common form of clause in an engineering contract, which concerns intellectual property, is an indemnity clause. This is likely to occur when one of the parties provides a design or a set of drawings or other documents to the other, from which items are to be made. In such circumstances it is possible that the design or drawings or documents might infringe the intellectual property rights of a third party, or that the manufactured product or the use of it may cause such an infringement. It is possible that the third party might bring an action against either, or both, the contractor or the purchaser. An indemnity clause would place the costs and the legal liability upon the party responsible for bringing about the situation, by creating the drawings, designs or other documents. Such clauses are usually drafted in very wide language, and usually cover all claims, expenses, costs, etc., arising from any infringement (or alleged infringement) of any patent, copyright, design, design rights, trade mark or other form of intellectual property protected by law. In cases where a form of contract is intended to become standard, and to be used and re-used over again, it is probably best to have two such clauses, one dealing with the situation where the purchaser provides the designs and other information which might infringe the intellectual property of a third party, and the other clause dealing with the situation where the design and information is provided by the contractor.

Table 8.2 Table showing different types of commercial contracts and the clauses about intellectual property most likely to be included in them

Type of clause \ Type of contract	Engineering and sales	Employment	Agency and distribution	Research/ authorship	Joint venture
Secrecy/confidence	Yes	Yes	Yes	Yes	Yes
Indemnity	Yes	No	Sometimes	Yes	Not usually
Allocation of property rights	Sometimes	Not common, but possible	Not usually	Yes	Yes
Agreement not to compete	Not usually	Sometimes	Yes	No	Yes
Grant of licence	Possibly	No	Sometimes	Yes	No
Royalty agreement	Not usually	No	Sometimes	Yes	Yes
Special clauses	Unusual	Unusual	Protection of rights	Moral rights	Allocation of rights after termination

Some legal questions answered

Is it true to say that if the parties make no express provision about the transfer of ownership in goods, the property always passes on delivery?

It is true in the majority of cases, but not true in all cases. The Sale of Goods Act 1979 instances at least four examples of situations where the transfer of the ownership of goods does not occur at the moment of delivery, but at some other time. This may seem like a rather large number of exceptions to the rule, but fortunately most of them do not apply to engineering contracts. An example would be a 'sale or return' contract, under which an item, for example, a machine tool, might be delivered to a purchaser who might wish to try it out before purchasing it. If this is the case, the implied term would be that title passes, not on delivery, but when the purchaser indicates that he has opted to keep the machine tool.

If risk goes hand in hand with the ownership of goods, does this mean that a seller retaining title to goods, or a purchaser acquiring title before delivery, must bear the risk?

Not necessarily. It is true that this would be the case under common law and under the basic provisions of the Sale of Goods Act 1979, but these provisions are only to be applied in the absence of any wording to the contrary in the contract itself. It is perfectly possible, therefore, to make express provision that the risk will be, for example, upon the purchaser immediately upon delivery to the purchaser, but the ownership will remain with the seller until the purchaser has paid in full all sums owing under the contract. A similar provision is possible, making changes where necessary, in cases where the purchaser wishes to acquire title to goods before delivery, while the seller is to bear the risk in the goods until delivery.

What is an 'all monies' clause, and how does it differ from any other type of Retention of Title clause?

A 'simple' Retention of Title clause is one which states that the seller remains owner of the goods until the purchaser has paid for *those goods* in full. There is nothing wrong with this type of clause provided that the seller can at all times trace the exact goods the subject of each invoice: it puts a premium on a high standard of traceability. If the item sold is a serial-numbered item, such as a machine or vehicle, this does not give rise to many problems, but if the items sold are less perfectly traceable, such as materials for manufacture, then sellers attempting to enforce the simple clause often find that receivers and liquidators challenge them to prove that the goods available for return are the same goods that match the unpaid invoices. On this point many a seller has had to admit defeat. The 'all monies', 'all sums', or 'all debts' clause is a variant upon the simple clause, which extends the Retention of Title until *all debts* owed by the purchaser to the seller, whether under the present contract or under any other contract, have been paid in full. The theory is that this will work even where the traceability of items delivered is inadequate to isolate one batch from another. As long as there is no doubt that the items of the same kind all come from the same supplier, and as long as all debts to that supplier have not yet been paid in full, that supplier should be able to recover such of its goods as are subject to the clause. Thus, in *Armour* v. *Thyssen Edelstahlwerke AG* (1990), a German steel producer was able to use such a clause to recover sheet steel which it had delivered to a purchaser which later went into receivership.

Is it necessary in the context of an engineering contract to state which party is entitled to patent any invention or to have copyright of any documents arising out of the contract?

This is really a matter of degree: some engineering contracts are of a fairly routine nature and relatively few documents in them are likely to contain any original material. It is unlikely that a new patentable invention will arise from such contracts. In these cases there is no real need for any special conditions allocating intellectual property rights between the parties. At the other end of the spectrum there are research and development contracts, in which the entire objective of a project will be to create new and valuable ideas and inventions, or processes or solutions to problems. The purchaser will be funding the project, and will usually expect to own the intellectual property rights that result from it. However, in some cases the contractor or consultant will be unwilling to give up all intellectual property rights, because to do so would be to deprive himself of the right to use or develop or re-cycle the ideas on other occasions for other clients. The outcome of this apparent conflict of interests will then be for the parties to negotiate.

Can any form of compromise be reached if both parties to a contract want some form of intellectual property arising from it?

Compromise arrangements are certainly possible. This can be done either by distinguishing between different things, allocating them according to agreement (for example, 'Document A will be your copyright; Document B will be our copyright'), or it can be done by one party having the intellectual property rights, and granting to the other a *licence* to do something which would otherwise infringe the intellectual property rights of the other. The licence can be limited by time and place, if necessary, and can either be royalty free, or in return for an agreed royalty.

What is the legal position if one person infringes the intellectual property rights of another?

There are several possibilities. The most likely one is an action for an injunction, which is a court order preventing the use or exploitation or publication of a document or other activity which is alleged to have infringed the intellectual property rights of another. If it is a party to the engineering contract who is seeking the injunction against the other, the grant or refusal of the injunction usually concludes the issue. However, if the injunction is sought by and obtained by a third party, then the terms of the engineering contract, and the relevant indemnity clause, will usually show where the costs and expenses and commercial loss are to fall. Apart from the remedy of injunction, there are other legal remedies available for an infringement of intellectual property rights, such as an action for damages, or an action for account of profits, which is a way of making the offending party pay the profits made out of the infringement to the owner of the intellectual property rights.

9 Multipartite arrangements

Agency, sub-contracting, and free-issue

In this part of this chapter some of the different types of arrangement and relationships that may be made in the course of engineering and other commercial contracts will be examined. The structure and nature of relationships is of great importance because it tells us about the following:

the allocation of obligations;
the level at which obligations are carried out;
the chain of instructions;
communications;
who bears what liability;
to whom liabilities and duties are owed.

Agency and sub-contracting

As an illustration of these relationships regarding agency and sub-contracting, one should look at the case of *Redler Grain Silos Ltd* v. *BICC Ltd* (1982). The facts of this case are complex, but for these purposes can be reduced to the issue that the purchaser of works in Iran wished to take over goods and materials which had been sold by BICC Ltd (the sub-contractor and manufacturer) to the main contractor, Redler Grain Silos. Doubtless the purchaser had in mind the possibility of dismissing the main contractor and having the work performed by another contractor, although we are not told the actual reasons for this. The question before the English courts was whether or not, assuming that it might wish to do so, BICC Ltd would be legally entitled to deliver the goods and materials, which had not yet been delivered under the contract, directly to the Iranian purchaser. If Redler Grain Silos had been merely acting as an *agent* for the Iranian purchaser in ordering the goods, then there is no doubt that BICC Ltd would have been entitled to deliver the goods directly to the purchaser. An agent is in law merely a channel or conduit through which the real contractual relationship between seller and purchaser is established. The agent acquires no property in the goods sold: but in this particular case the court held that the structure of the relationship was one of *two separate contracts*, under one of which BICC Ltd was to sell goods to Redler Grain Silos, and under the other one of which Redler Grain Silos was to put up works for the purchaser. Consequently, there was no direct contractual relationship between BICC Ltd and the purchaser, and BICC Ltd would be in breach of its contract to Redler Grain Silos if it were to make direct delivery to the purchaser.

This, of course, is not to suggest that there can never be any direct delivery by a sub-contractor or supplier to the end-user: such a suggestion would be contrary to commercial reality. What is, however, true is that before we can appreciate who is entitled to what, we must look at the full set of contracts and the relationships established by them. The same is true of direct payment by an end-user to a sub-contractor or supplier, without routing the payment through the main contractor. In the absence of any agreed and legally valid terms permitting such a procedure (the main contractor being party to such an agreement), it would not be legally permissible. It can and does occur, but only where the terms of the contracts facilitate it. In Figure 9.1 we can see the difference between *agency* and *sub-contracting* from the point of view of legal structure. Apart from the point about property rights and the rights to delivery and payment, there are further legal differences to be noted.

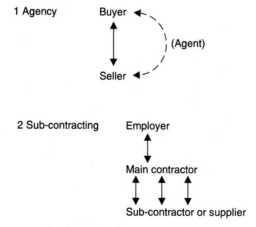

Figure 9.1 *Agency and sub-contracting compared*

Instructions

If work is sub-contracted, instructions must always be from the purchaser to the main contractor, who may then pass them on to the sub-contractor. If the purchaser or the purchaser's engineer were to instruct the sub-contractor directly, this would potentially be an obstruction of the main contractor's work and could have many different consequences; for example, releasing the main contractor from liability for that particular instruction, or a possible liability on the part of the purchaser to pay damages to the main contractor if there were adverse financial consequences.

Warranties and defects liability

As we have already noted in an earlier chapter, warranties and defects liability are the result of a contractual relationship between two parties, so the normal position will be that the manufacturer of components will warrant them to the contractor and the contractor will in turn warrant the entire works to the end-user. A direct warranty will not usually exist between the manufacturer of components and the end-user of works, unless either there is a *collateral warranty* between them, or unless the goods have been purchased by a contractor acting as *agent* for the end-user.

Agency

The nature of the agency relationship can easily be seen from Figure 9.1. The purchaser who uses an agent is, in law, buying directly from the source (seller or manufacturer). The agent is not truly a party to the contract, but only a means by which the contract is made. The result of this is that liabilities and warranties are exactly the same as in any other supply contract. This point arose in the case of *Teheran-Europe and Co. Ltd v. S. T. Belton (Tractors) Ltd* (1968). This case was about two important and related issues. One was whether or not the seller was in breach of the implied term as to their intended purpose. The other issue was whether or not Teheran-Europe and Co. Ltd, who had purchased the goods through an agent, were entitled to make a direct claim against the sellers. The goods consisted of twelve air compressors which had been forwarded to Iran, but which, unfortunately, did not comply with local regulations. The Court of Appeal found in favour of the buyer on the second point: a buyer through an agent is entitled to claim directly against the seller with whom the buyer, in fact, has a contract. However, the court found against the buyer on the question of fitness for purpose: the case is a classic illustration of the principle which we have already encountered, that it is up to the buyer to make known to the seller the particular purpose for which the goods are required, and since the English seller had not been informed of the content of the local regulations, or even been told of a requirement to comply with them, the seller could not be liable for any failure of the goods to meet that particular purpose.

Free-issue

The expression 'free-issue' is so new to the law that it does not, as yet, have an established meaning. However, it has been used for many years in commerce, in particular in engineering contracts. It needs to be discussed at this point because it is different, in the structure of the relationships and legal liabilities created, from either the agency or the sub-contractor/main contractor/purchaser relationship.

Figure 9.2 *The 'free-issue' relationship*

As can be seen from Figure 9.2, the free-issue relationships come into play when the sale of manufactured goods or materials is by a third party, such as a manufacturer directly to the employer or end-user. The employer will then issue these goods to the contractor, either for incorporation into the works, or for use during the course of the works. From the employer's point of view, there is often a commercial advantage to be gained by a free-issue of goods: it may be that the employer will be able to take advantage of its superior purchasing power, and thus obtain better prices than the contractor could obtain; or it may be that the employer will be able to achieve a requirement of standardization which might otherwise be difficult to obtain, by buying and free-issuing its own goods and materials. Furthermore, the problem of how to create the 'warranty' relationship directly with a manufacturer of goods is overcome by using the free-issue method of obtaining goods.

However, there is a price to pay for the advantages obtained by the free-issue method. The employer will not be able to look to the contractor for every aspect of defects liability, as is normally the case, because the free-issue relationship will distort this liability to some extent. In some cases, far from the employer being the one to complain to the contractor about defective goods and materials, it will be the contractor who is complaining to the employer about additional costs, or about delays, caused by defects in free-issued goods. In Canada, this point was illustrated by the case of *Perini Pacific Ltd* v. *Greater Vancouver Sewerage and Drainage District* (1966) in which the delivery by the employer to the contractor of defective machinery, on a free-issue basis, meant that the employer was to blame for the resulting delay. In normal circumstances, a delay due to defective equipment or components will be the liability of the contractor, who will be exposed to liquidated damages, but this was not the case where the defective items were free-issued.

Good engineering contracts should therefore take account of the possibility of free-issues, and should state which items are to be free-issued, at what time, and at what place. There should also be a procedure for the contractor to report defects in the free-issued items, so as to minimize any delay.

The chain of responsibility

Leaving aside agency and free-issue, the most common structure of an engineering contract is one in which responsibilities and liabilities flow down in a chain from the employer to the contractor, and from the contractor to sub-contractors, and from sub-contractors to sub-sub-contractors and suppliers. The result of this should be that if, for example, a component in a structure proves to be defective at some point after delivery and taking-over, the employer can instruct the main contractor to repair or to replace the item in question, and further instructions will be passed down the chain of responsibility, until the originator of the item either repairs or replaces it or bears the cost of doing so. The legal point about the chain of responsibility is that the main contractor remains fully responsible to the employer, and cannot step aside or refuse to accept liability. If, for example, the supplier of the defective component were to cease to manufacture, or to go out of business, it would be up to the main contractor to find a way of coping with the defect; the question of whether or not the main contractor can then pass back the costs of putting right the defect is a risk that the main contractor would have had to bear in mind at the time of making the relevant contracts. It follows that main contractors, and others who find themselves in the middle of a chain of responsibility, must look at their contracts of purchase, as well as at the sales contract, so as to make sure that the duties, liabilities and responsibilities are compatible.

'Back-to-back' contracts

The expression 'back-to-back' is often used to mean compatibility of terms in a chain of contracts in an engineering project. There is nothing wrong with the expression, provided that it is understood that there is no official or standard meaning of 'back-to-back', and that its practical application can differ a great deal from project to to project. The terms in a chain of contracts are unlikely to be identical, because the functions of various parties are completely different, and the timing within a programme will be different, as will the prices and values of work done. What is important is not that terms in a chain of contracts should be identical, but that they should be compatible, and from a main contractor's point of view, they should allow the main contractor to recoup any losses caused by delays or defects for which sub-contractors or other parties are responsible.

Breaks in the chain

All parties to a chain of engineering contracts need to bear in mind the possibility that the chain may become 'broken' in the course of events, leaving a liability which cannot be passed on. This fact can be seen from some of the cases encountered in Chapter 7, and understanding and coping with it is part of the techniques of risk analysis in an engineering project.

Chains of responsibility can become broken for a number of possible reasons. Insolvency of one of the parties is one of the most likely reasons, and the possibility of it is one of the commercial factors in deciding whether or not to require a performance bond or other form of guarantee for due performance. One of the objectives of such a bond or guarantee is to allow a purchaser or main contractor to recoup the costs of having to have work or remedial work done by another party if the contractor or sub-contractor from whom the original services have been purchased becomes insolvent.

Insolvency is probably the most common cause of a break in a chain of contractual liabilities, but it is by no means the only one. Another less well-known cause is inequality or incompatibility of terms. A main contractor may have given a warranty which begins in the year 1995, and which expires in 1997. His sub-contractor may have given a warranty of exactly equal length, but which starts and expires in 1994 and 1995 respectively. Superficially, the two contracts appear to be 'back-to-back', but in reality the liabilities occupy different periods of time, and are not fully compatible.

Yet another example of a possible break in the chain will occur if a contractor purchases components on more onerous terms than the terms of the main contract: there may, for example, be an exclusion of liability in the sub-contract, but none in the main contract. Some exclusions of liability (such as agreed exclusions of 'consequential' loss or damage, as defined by the contract) are capable of standing up in law. If this occurs, the main contractor will be left holding a loss or liability which it is excluded from passing on to the party causing it. It was for reasons of this kind that negotiations became so fraught with difficulties in the case of *British Steel Corp. v. Cleveland Bridge and Engineering Co. Ltd*, which is discussed at considerable length in Chapter 1 of this book. CBE was, at the time of negotiating with BSC, already committed to a main contract which imposed severe liability for delay. BSC were willing to contract to supply components within the agreed timescale, but were unwilling to accept terms which passed down to them a liability in respect of any delay in delivery. BSC offered terms of sale which excluded liability for 'consequential' loss, which was presumably defined in such terms as to effectively disentitle CBE from recouping any loss caused by delay on the part of BSC. Negotiations broke down, and a contract was never made. One of the reasons for this failure must surely have been the inability of the parties to 'square the circle', that is to say, to find a way of managing the risks of delay in terms which would have been acceptable to both parties. This is surprising, when it is borne in mind that there would have been a number of possible ways of doing this. One technique would have been a compromise on liquidated damages, whereby BSC would have agreed to accept a scale of liquidated damages for delay which allowed CBE to recoup some, but not all, of any possible liability to the employer for delay. This technique of spreading the risk is often used in chains of engineering contracts, bearing in mind that some sub-contracts or sub-sub-contracts will often be of relatively low value compared to the main contract. Another possible way of dealing with the problem would have been for CBE to have recognized that it was in a position in which it stood to lose the most by any failure to conclude a contract. In such a position, a main contractor may find that due to lack of bargaining power, it is simply not able to obtain the kind of 'back-to-back' terms that it would like. In these circumstances one must recognize realities, and either insure against liability for delay, or be prepared to pay a premium for priority treatment from the supplier.

Nominated sub-contractors

The possibility that a sub-contractor may be nominated by the employer or purchaser raises further questions about the chain of responsibility. The legal and commercial concept of the nominated sub-contractor occurs when the purchaser instructs the main contractor to use a particular person or company as a sub-contractor in respect of all or part of the works. There can also be nominated suppliers, for the supply parts of contracts. The gain to the purchaser from nomination is similar to that arising from free-issue: a possible price advantage, and standardization with goods and services already owned by, or given to, the purchaser. Nomination may indeed occur for purely commercial reasons, such as the track record of, or previous association with, the nominated sub-contractor.

The most likely question to be raised by a main contractor about nominated sub-contractors is as to whether or not the very fact of the sub-contractor being the purchaser's own choice causes a 'break' in the chain of responsibility, and relieves the main contractor from liability for delay or defects caused by the nominated sub-contractor. It is, at first sight, an attractive argument, but in practice the courts have shied away from it. The courts have to take a path of commercial good sense and convenience. If, in every case where there was a nominated sub-contractor there was *also a collateral contract* between the purchaser and the nominated sub-contractor, then the courts might have followed the line of argument put forward by some main contractors wishing to be relieved of liability. However, in many cases there is no such collateral contract, and the absence of it would leave the purchaser without any recourse unless the courts were to hold that the main contractor remains responsible. So, by and large (subject to one or two possible exceptions), that is the position: the fact of nomination makes no difference, and the chain of responsibility holds good. Lord Reid summed this up, and explained an exception, in the case of *Young & Marten Ltd* v. *McManus Childs Ltd* (1969). In this case the manufacturer of tiles was a nominated supplier, and the main contractor hoped to use this fact to avoid liability for defects in the tiles. Lord Reid stated:

> Why should that make any difference? It would make a difference if the manufacturer was only willing to sell on terms which excluded or limited his ordinary liability under the Sale of Goods Act and that fact was known to the employer and the contractor when they made their contract.

The second sentence of this passage shows that exceptional cases involving nominated sub-contractors or nominated suppliers can exist; Lord Reid hints at one of them, but doubtless there may be other exceptions as well. The one pointed out by Lord Reid serves as a warning to purchasers and employers: if it is known at the time of the nomination that the nominated sub-contractor has onerous terms giving the main contractor no recourse, then if the employer persists with the nomination, this may imply that the employer accepts the risks of using such a nominated sub-contractor, and is prepared to relieve the main contractor of liability.

Intervening circumstances or acts of a third party

Once the legal framework and structure of relationships in an engineering project are in place, nothing should be done to interfere with or disturb the position. If a problem, whether of functionality or of safety arises, the employer should refer this to the main contractor, whose responsibility it will be – whether under a continuing contract, under a warranty, or as part of the employer's statutory rights. If the employer approaches a different person, such as a sub-contractor or a third party, this may cause legal complications of the kind which arose in the recent case of *Beoco Ltd* v. *Alfa Laval Co. Ltd and Another* (1994). In this case Alfa Laval Co. Ltd had been responsible for installing a heat exchanger at the works of Beoco Ltd. A crack appeared in the exchanger in August 1988, and in October 1988 it exploded causing damage to equipment and lost production. In these circumstances it is reasonably clear that the seller or contractor has a liability both to repair or replace the defective part of the machine, and to pay damages for lost production. This would have been the position of Alfa Laval Co. Ltd, but for one particular intervening factor. On 24 August 1988, another engineering company repaired *part* of the crack which had appeared in the heat exchanger. This in itself was insufficient to repair the full extent of the defects. The court found subsequently that Beoco Ltd had failed to make sufficient tests to ensure that the crack had been properly repaired, and this was apparently the reason for the explosion. The Court of Appeal allowed the appeal of Alfa Laval against the quantum of damages, and held that the damages to be awarded should be limited to the cost of replacement of the defective casing of the heat exchanger and *only* those losses of production that took place while the repair was being effected. Other losses of production after the explosion were not caused by Alfa Laval, but by the repair by the other contractor and by the conduct of the purchaser.

With hindsight, in a case such as this, the employer/purchaser, after noticing the defect, should place complete responsibility for the repair or replacement and inspection upon the supplier or main contractor.

Some legal questions answered

What is the difference between an agent and a distributor?

The answer to this question is complicated by the fact that the word 'agent' is often used incorrectly in commerce. We tend to speak of a company being an 'agent' for certain types of products, such as machine tools or motor vehicles. What is usually meant by this is that the company in question is a *distributor*. An authorized distributor promotes and sells goods of a certain type and certain makes; but the distributor is not an agent in the true legal sense of the word. The true agent always buys or sells on behalf of another person. The distributor first of all buys goods from the manufacturer, and then, in separate legal transactions, sells them on to other parties. The distribution agreement is concerned with the method of promotion and sales, advertising, retaining goodwill, etc., as well as with such matters as areas and exclusivity.

When are agents used in engineering contracts, and why?

The most obvious example of an agent in an engineering contract is the engineer. The engineer, if appointed, is the representative of the employer. In law, and under the terms of the contract, this appointment confers an agency for certain purposes. What these purposes are depends upon the contract, but usually they include the supervision and certification of work, and the ordering of variations. As the engineer is an agent for the purposes stated, any action taken by the engineer within his authority is, in law, action taken for and on behalf of the employer.

Apart from the engineer, there are many other examples of the use of agents in engineering contracts. They may be used to obtain and facilitate services with which the parties have no experience of their own (shipping services, obtained via shipping agents, for example); or they may be used because of unfamiliarity with local conditions abroad; or because local laws actually require the use of local agents in some parts of the world.

How does one make distinctions between different types of sub-contractors?

The distinctions are really distinctions of definition rather than of law. It is up to the parties to an engineering contract to decide, firstly, whether or not there is to be complete freedom on the part of a main contractor to choose his sub-contractors, and, secondly, if the freedom is to be restricted, how the restrictions are to operate under the terms of the contract. The categories of sub-contractor are simply shorthand ways of referring to these matters. Thus, the expression 'domestic' sub-contractor means simply that the main contractor has been given complete freedom to choose a particular sub-contractor. The expression 'nominated' sub-contractor means that the main contractor has been instructed by the employer (or employer's representative) to contract with a particular sub-contractor for certain goods or services. These two expressions can be elaborated by the definitions contained in the contract. In between the two ends of the spectrum there are also a number of other possibilities, which differ as a matter of degree from each other. An 'approved' sub-contractor, for example, is basically the same as any other domestic sub-contractor, the only difference being that the main contractor must first obtain approval from the employer, or must select from a list of approved sub-contractors provided by the employer. A 'named' sub-contractor is very close to the concept of nominated sub-contractor, but may differ in so far as the 'name' put forward may not be a strict instruction by the employer, and may permit negotiation by the main contractor of alternative choices of sub-contractor.

What rights has a main contractor to object to sub-contractors nominated by an employer?

This will depend entirely upon the terms of the contract; many good standard forms of contract do in fact provide for some form of reasonable objection on the part of the main contractor. This is not a strict requirement of law, however, and many purchaser's forms of contract do not allow for any objection to be made. The existence of nominated sub-contractors, together with no right to make reasonable objection to any particular nomination, would be the kind of risk which the main contractor would have to consider carefully in this area.

If a sub-contractor becomes insolvent, what is the legal position?

The answer to this question shows that different ways of appointing sub-contractors may give rise to different legal consequences. If a sub-contractor is selected by the main contractor, then the risk falls entirely upon the main contractor. This means that any delays or wasted expenses or costs involved in having to select another sub-contractor will fall upon the main contractor. If the sub-contractor is nominated, there may be differences in the position, and these will depend upon the terms of the contract. Much of the risk will still fall upon the main contractor, because the terms of the contract, the system of payment, and the other commercial issues, will be for the main contractor to take into account when forming the contract with the sub-contractor. However, it may, under the terms of the particular contract, be the employer's duty (or the engineer's duty) to take action to deal with the circumstances following the insolvency of a nominated sub-contractor. Such action could involve the employer in having to nominate a replacement sub-contractor; or it could involve the engineer in having to make a variation order – either to vary work, or to omit it, or to have it done by the main contractor himself.

Can a sub-contractor validly make a claim for the recovery of goods sold by a main contractor to an employer?

This is unlikely to occur unless either the main contractor becomes insolvent, or there is a breach of contract by the main contractor, which entitles the sub-contractor to recover any goods or materials. We have already seen in Chapter 8, in the *Dawber Williamson* case, that it is possible for a sub-contractor to recover its own goods from site. The essential point in that case was that Dawber Williamson had not sold the tiles, but had delivered the goods to site to be fixed by themselves. However, the legal position may be different if the contract is one of *sale*. In such circumstances a triangular relationship is often set up, whereby the sub-contractor contracts with the main contractor to sell goods (such as items of machinery) to the main contractor, and to deliver those goods to the premises of the employer. In the intervening time before delivery, the main contractor will in turn, perhaps, have sold those goods by making a contract of sale to the employer. If this occurs, then instead of a chain of building or construction contracts, there is a chain of sale of goods contracts. At first sight, the difference may appear to be of little consequence, but for the reasons already mentioned, there is, in fact, a significant difference as regards title to goods. Instead of the parties being assumed to be working on site with their own goods, there will be sales

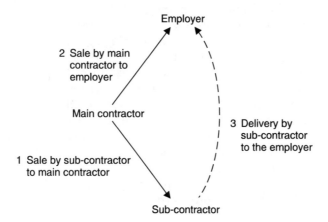

Note: 1 The above sequence of events will confer title to the goods upon the employer.
2 If the main contractor subsequently becomes insolvent, the employer will keep title to the goods, but the sub-contractor will have to look to the liquidator of the main contractor for payment.

Figure 9.3 *The insolvency of a main contractor*

150 Multipartite arrangements

under which ownership of the goods will pass immediately upon delivery, in the absence of any provisions to the contrary.

Figure 9.3 shows a hypothetical position, where a sub-contractor may have sold goods to a main contractor, and may have delivered those goods to the site of the employer. The main contractor may, in turn, have sold those goods to the employer, and may subsequently have become insolvent. The sub-contractor, in these circumstances, will probably not have received payment and may approach the employer, requiring either the return of the goods or payment. In reality, there is no legal obligation on the part of the employer to do either of these things. The chances are that the goods will have become the legal property of the employer by virtue of delivery. As for payment, the terms of the contracts, together with the laws of insolvency, will normally prevent the employer from making any payment to the sub-contractor in these circumstances. Payment will have to be made to the receiver, or administrator, or liquidator of the main contractor, and the sub-contractor will have to claim as an unsecured creditor from them.

The only way in which a seller of goods, which are to be sold on to the end-user, can make sure of keeping title to the goods in such circumstances is to try to ensure that two things occur. The seller must have a good and effective Retention of Title clause in his conditions of sale to the main contractor: but this alone is not enough; the clause will not bind the end-user. So the original seller must also insist that the main contractor uses a similar Retention of Title clause when the main contractor sells on to the end-user. This is, in practice, very difficult to achieve, but recent cases show that from a legal point of view it is effective. It worked in *Hanson W. (Harrow) Ltd* v. *Rapid Civil Engineering Ltd* (1987). More recently, this view was confirmed in the case of *Re Highway Foods International Ltd (in administrative receivership)* (1994). This case was not an engineering case, but its logic would apply to any sales of goods in a chain of sales. A company called Harris Ltd supplied meat to Highway Foods Ltd, on Retention of Title terms. Highway then sold it on to another company, Kingfry Meatproducts Ltd. Highway became insolvent, and title to the goods became the issue. The High Court held that since the sale by Highway to Kingfry also included a Retention of Title clause, title remained with Harris.

What are the most significant practical difficulties in the normal structure of main contracts and sub-contracts?

Apart from the *legal* problems, about title and about liabilities and warranties, the main *practical* problems are ones of effective communications. The longer the chain of contracts, the more difficult communications may become, since the strict legal requirements may collide with practical solutions to problems. An employer may find that queries of a technical nature have to be routed to a sub-sub-contractor down the chain, and that (in terms of speed and effectiveness) this simply does not solve the problems as and when they arise; but to communicate with the sub-sub-contractor directly can only confuse the legal issues.

These problems regularly confront businesses, and it is not easy to find ready-made solutions to them. One possibility already discussed is the free-issue route to supply of a particular item. This will work where the sub-sub-contract concerns specific goods, such as items of plant or machinery, or hardware or software. It is less easy to apply where the sub-sub-contract consists of a design or work or services element. Another possibility is for the employer to contract directly with the sub-sub-contractors and to employ a management contractor to manage and supervise the project. This, in effect, means that there are no sub-contractors or sub-sub-contractors, as they will all be in direct contractual arrangements with the end-user. This method has been tried on many large projects, and in theory it should provide solutions to the problems mentioned – but it can also bring its own peculiar difficulties and ambiguities, and should only be used after considerable thought by those who have considerable experience in handling such contracts.

10 Negotiating legal and financial matters

Commercial and engineering contracts are complex organisms, with internal structures analogous to those that occur in biology or in chemistry. This book has looked at different aspects of contracts, separately, but occasionally relating them to one another with a forward or a backward glance. In this final Chapter, we look at some of the remaining legal fibres that hold the contractual structure in place and which help to prevent it from collapsing or flying apart under its own stresses. This chapter will look in particular at performance bonds and guarantees, at certain aspects of insurance, and at methods of dispute resolution. One factor that all of these types of provision have in common, and which distinguishes them from conditions about price or performance or quality, is that they all deal with contingencies which are unlikely to occur in the majority of cases. Bonds and guarantees are 'called' infrequently. The intention of the contractor is that they should never be called. Insurable risks materialize only in a minority of cases. The arbitration or jurisdiction clauses in an engineering contract are used so infrequently that they tend to be overlooked: a dangerous tendency in view of the problems this practice caused to the parties in the *Brinkibon* case, mentioned in Chapter 1 of this book.

Performance bonds and guarantees

The main purpose of bonds and guarantees is to provide some form of security for the purchaser. A certain linguistic problem arises because of technical differences in form and in purpose, and because of different usage of words in different areas of commerce. To put the question simply, readers will no doubt wish to know whether or not there is a discernible difference between a 'bond' and a 'guarantee', and if there is one, what that difference is. Before tackling this point, it must be made clear that English law has always been more concerned with intention and with content, as discerned from the precise wording of a document, than it has been with headings or purely descriptive words. English law tends to look to substance rather than to form. For this reason we need to look at the commercial nature of bonds and guarantees (taking the two words to mean substantially the same thing), before discussing the shades of legal distinction that they may give rise to in their different forms.

Different types of bonds for different purposes

Bonds may be classified according to *function*, as follows:

Tender bonds (sometimes known as bid bonds)

These are bonds which are to be provided, if required, at the tender stage of a proposed contract. By no means all tenders involve such a process, but a number of tenders do, particularly larger international tenders. A tender bond is usually only for a small fraction of the likely contract value, such as one per cent or two per cent, and should never be for more than five per cent at the highest. It is inevitably a financial risk for the contractor, who has, at this stage, no certainty of any reward. The justification for it is that it provides the purchaser with a number of advantages. It enables the purchaser to ascertain the position of intending contractors, particularly as regards the ability to provide the other forms of bond which are likely to be required under the contract. It makes it most unlikely that the tendering contractor will withdraw, or attempt to amend, the tender after it has been made, and before acceptance, thus counteracting the rigours of the common law, which permit the withdrawal of any offer at any time before acceptance. It also provides the purchaser with a source of funds, if, following expensive negotiations, the successful tenderer refuses to enter into a contract with the purchaser. This situation can come about because although the purchaser can turn a tender into a contract by accepting it not every intending purchaser does this, and often the negotiations constitute *counter-offers*, which destroy the legal character of the tender as an offer.

Advance payment bonds (sometimes known as down payment bonds)

These are intended as a form of security for the purchaser in respect of advances made to the contractor. The purchaser who makes an advance or 'down payment' will normally be taking a financial risk, since no value by way of goods or services will as yet have been received. It has already been shown that early acquisition of title to any goods or materials intended for the contract is desirable, but even this will not always be secure, since the goods might not even exist or be identifiable at the relevant time. The advance payment bond should enable the purchaser to receive a refund of such money even if the contractor becomes insolvent. The value of such bonds depends upon the wishes of the parties: it may match the advance payments or be a fraction of them; it may rise under the terms of the contract, or it may reduce, as and when payments would normally be due for value received; or the contract may provide for a more complex system in which the value of the advance payment bond is first stepped up then stepped down.

Performance bonds

These are the 'classic' bonds, which are intended as security for the performance of the contractor's obligations. They are usually provided at the time of the entering into the engineering contract or, alternatively, within a stated time of so doing (for example seven days). The amount of such bonds is one which varies a great deal, depending upon the negotiations, but the norms are between five and fifteen per cent of the contract value. The aim is to give security to the purchaser as to the contractor's financial standing, and also to provide a source of compensation to the purchaser if the contractor should default. In particular, the bond is intended to provide ready money for payment of another contractor, if it becomes necessary to replace a defaulting contractor.

Bonds in lieu of retention (sometimes known as warranty bonds)

Retention money was looked at in an earlier chapter of this book. Its main function is to ensure that the contractor will perform remedial work properly, and that failing this, the purchaser will be able to use the money to have the remedial work done by a different contractor. A bond of sufficient value can fulfil much the same function, although the purchaser will tend to prefer the retention of money, since it is unfettered cash (and is probably easing the purchaser's cash flow or earning him interest). However, the system of the purchaser retaining sums of money earned by the contractor, often as high as ten per cent, can be most disadvantageous for the contractor, who cannot realize his profit, and loses cash flow for a period of between six months and a year. A bond with suitable wording which enables the purchaser to obtain the money easily on demand is a compromise, if exchanged for all or part of the retention money.

How bonds work

The form of a bond or guarantee will usually be set out as a term of the engineering contract between the purchaser and the contractor, but the bond itself will be a separate contract. Thus, an engineering contract may state that within seven days of the making of the contract the contractor will provide the bond or guarantee, from a bank or insurance company or another agreed surety, in the form set out in the appendix to the contract. This, if the engineering contract is agreed, is a material obligation, and failure to carry it out may result in termination of the contract. Those negotiating engineering contracts should never enter into any engineering contract which provides for a bond or guarantee unless its form and source is fully known and also agreed. Disputes can arise, if this advice is not headed, as to what is an acceptable bond and what is not.

Once the bonding requirements are known, the contractor will make arrangements with his bank or insurance company. The bank or insurance company may well require an *indemnity* from the contractor, so that any calls on the bond will ultimately be paid for by the contractor, when the indemnity is enforced. But the bond stands quite independently of the indemnity or of the finances of the contractor. The bond is a direct undertaking by the bank, insurance company or other surety to the purchaser to pay a specified sum of money to the purchaser. It may be payable on demand, or payable subject to the satisfaction of certain conditions. The contractor will have to pay a fee or setting-up charge to the bank or insurance company in return for the facility represented by the bond: in the case of a bank, this facility is equivalent to an overdraft, and indeed some contractors will prefer to use insurance companies as providers of bonds, so as not to affect their bank overdraft.

If the purchaser so requires, particularly if the purchaser is an overseas company or body, the bond may be given by the contractor's bank to the purchaser's own bank. In this way, a chain of undertakings may be set up, as illustrated by Figure 10.1.

Note: The unbroken arrows represent the bonds or undertakings given; the broken arrows represent the calls or demands that may be made.

Figure 10.1 *Bonds: a chain of undertakings*

The legal basis of the bond

Bonds work on a contractual basis. The contract is an independent one between the bank or other surety and the purchaser. It depends upon its own conditions. One of the questions that sometimes arises is the *consideration* that is given by the purchaser in return for the undertaking given by the bank. The law on this is far from certain, particularly when bonds are given in overseas contracts, but those wishing to be certain in these matters should takes steps to ensure that the bond either expresses a form of good consideration, or alternatively that it is executed under seal, as a deed (in which case no consideration is required). Both methods require great care and advice from specialists, to make sure that the execution as a deed, or the consideration, is properly expressed.

Bonds as compared with Parent Company Guarantees

Bonds are capable of being written in either the 'on demand' form, or in a conditional form. If conditional, the bond will contain words which describe the necessary conditions for either the discharge of the bond or for the payment of money under the bond. If, on the other hand, the bond is an 'on demand' bond, then the bank or other surety has no duty to investigate the truth of any matters alleged, but is only concerned to ensure that the demand is made in the proper form, accompanied by the required documents.

A Parent Company Guarantee may be agreed upon by the parties, instead of a performance bond given by a bank or insurance company. Again, its exact legal nature will depend strictly upon its wording, but in practice such guarantees are usually conditional in form. They are usually guarantees that if the contractor, being a company wholly owned by the parent company, fails to carry out and complete its obligations under the contract, the parent company will be answerable for the discharge of such obligations.

The advantage of opting for Parent Company Guarantees to the contractor (and to the parent company, whose consent is, of course, required) is the saving of costs and the freeing of bank facilities which would otherwise be tied up by bonds. This can be particularly useful where the contractor has several contracts running at the same time and is faced with a number of simultaneous requirements for bonds. There may, however, be internal or policy reasons why some parent companies will not be prepared to guarantee the performance of obligations by their subsidiaries. At first sight there does not appear to be any obvious advantage to the purchaser in accepting a Parent Company Guarantee in lieu of a bond, but commercial, as compared with strictly legal, considerations are likely to be the decisive factor.

'Letters of comfort' as compared with guarantees

A Letter of Comfort should not be mistaken for a guarantee: it is most likely to be offered when a guarantee has been asked for, and when a third party, such as a parent company does not wish to guarantee the performance of its subsidiary. This fact may be made clear by a covering letter, or it may be obvious from the fact that the Letter of Comfort is quite different in wording from a guarantee. A Letter of Comfort is sometimes rather like a reference, describing the track record of the contractor, but never at any time guaranteeing, or warranting, or undertaking any form of liability for the performance of the contract by the contractor. Sometimes the Letter of Comfort is even more non-committal than this; it may be ambiguous in its wording, in which case legal advice should be sought as to its meaning and effect. It is one thing to accept a Letter of Comfort knowing that it has only commercial value, but no legal value at all; it is another thing altogether to accept it, thinking that it is as good as a guarantee.

In *Kleinwort Benson Ltd* v. *Malaysian Mining Corp. Berhad* (1989), the Court of Appeal had to consider the legal effect of a letter which stated (in response to a request for a guarantee by a parent company): 'It is our policy to ensure that the business of MMC Metals Ltd is conducted in such a way that MMC Metals Ltd is at all times in a position to meet its liabilities to you under the above arrangements.' After this letter had been issued by the parent company to Kleinwort Benson Ltd, Kleinwort Benson Ltd made loans amounting to £10 million to MMC Metals Ltd, a wholly owned subsidiary of Malaysian Mining Corporation Berhad. MMC Metals Ltd then ceased trading, and Kleinwort Benson Ltd looked to the parent company for repayment of the loans, together with interest. As it was clear that the word 'guarantee' had not been used, nor any words indicating any form of indemnity, the only possible basis for a claim by Kleinwort Benson Ltd was that the letter amounted to a form of *contractual warranty*. This approach found favour with the judge of the High Court, but on appeal to the Court of Appeal the decision was reversed in favour of Malaysian Mining Corporation Berhad. The Court of Appeal noted that no words such as 'promise', 'warrant' or 'undertake' had been used, and found it difficult to construe a warranty in what was a statement of *present fact* rather than an undertaking to maintain that state of affairs at all times in the future until the loan was repaid. The words 'It is our policy' were not the same as 'We undertake that it will be and will remain at all relevant times our policy'. However, the mystery is why the letter was thought to be adequate in the first place. If a genuine guarantee is required, it should be insisted upon!

What are the legal implications of bonds and guarantees?

At this point we have to tackle the linguistic difficulty hinted at earlier in this chapter. The law does not always follow the same practical thought-processes as people in commerce. As far as those in industry and commerce are concerned, all that really matters is the value of the bond, its timing and period of validity, the identity of the bond-giver or surety, and whether it is payable on demand, or on conditions which have to exist before payment. In an ideal world, those involved with engineering contracts should be concerned only with these straightforward issues, and not with the historic or linguistic problems that surround the words 'bond' and 'guarantee'. Having said this, it is reasonably clear that a Parent Company Guarantee is not identical to a bond, and the very word 'guarantee' is capable of a number of possible meanings in English law. For this reason, it is probable that shades of legal distinction can be found between the implications of guarantees and bonds, even though the difference is more likely to be found in the *content* of the document rather than the title or heading.

In the case of *Edward Owen Engineering Ltd* v. *Barclays Bank International Ltd and Umma Bank* (1978), a contract had been made between an English company and Libyan customers for the supply of glasshouses. Payment was to be by an irrevocable, confirmed letter of credit. The Libyans had required a performance bond, payable on demand without proof or condition. This was issued by the Umma Bank to the Libyan customer, and Barclays Bank International Ltd gave a bond in similar terms to the Umma Bank. After the contracts had been made, Edward Owen Engineering Ltd was not satisfied with the letter of credit, in the form offered, since it was not confirmed as required. The company therefore repudiated its obligations under the engineering contract. The response of the Libyan customer was to make a call on the performance bond. The Umma bank then made a call, in turn, on the bond issued by Barclays Bank International Ltd. Edward Owen Engineering Ltd sought an injunction in the English courts to stop Barclays from making payment on the bond. The injunction was not granted. Both the High Court and the Court of Appeal held that the bank should not be concerned with the rights and wrongs of the situation between the contractor and its customer, but had simply to pay on demand. The only remedy for the contractor was to bring a separate action against its customer for damages for breach of contract, under the terms of the engineering contract. (Unfortunately for the contractor, this particular remedy would probably have been difficult to enforce against a foreign customer, outside the jurisdiction, with a contract which was probably subject to local law and jurisdiction in the customer's own country.) This case illustrates the realities of bonds, particularly in international dealings, and particularly if the bond is in the 'on demand' form.

However, the recent case of *Trafalgar House Construction (Regions) Ltd* v. *General Surety & Guarantee Co. Ltd* (1995), shows that the general principle stated in the Edward Owen Engineering case is not applicable to all forms of bond, and may, in fact, only be applicable to the 'on demand' type of bond. With the 'on demand' bond, it may generally be said that a demand stated to be on the basis of the event specified in the bond is sufficient to activate the liability of the surety to pay. Many types of bonds issued in the United Kingdom are not, however, of the 'on demand' type. Many are based upon specimens provided by national institutions such as the ICE (Institution of Civil Engineers). Such bonds are *conditional* in the sense that the bond is stated to be null and void 'if the contractor shall duly perform and observe all the terms provisions and conditions of the said contract'.

The question which arose in the recent Trafalgar House case was whether, if such a conditional bond was 'called' by the holder, payment could be opposed by any claims or counterclaims or rights of set-off which the contractor might have against the holder of the bond. In this case, the holder of the bond was, in fact, a main contractor, and the 'contractor' was a sub-contractor. The sub-contractor was unable to carry out the work due to administrative receivership. Trafalgar House completed the work itself, and made claim under the performance bond. The giver of the bond, General Surety & Guarantee Co. Ltd, defended the case by raising questions of set-off, sums due to the sub-contractor, and cross-claims. In the High Court, it was held that such defences could not be raised if a valid demand on a bond had been made. This view was confirmed by the Court of Appeal, but the House of Lords held unanimously that with this particular type of *conditional bond*, such a defence was possible. General Surery & Guarantee Co. Ltd was given leave to defend.

What this means is that for legal and practical purposes there is a definite distinction to be made between the 'on demand' type of bond and certain types of conditional bonds. Whether this applies to all types of conditional bond remains to be seen, but the type which can only be 'called' if the contractor (or sub-contractor where relevant) fails to perform the work properly, and which otherwise is null and void is, in law, a *bond in the nature of a guarantee* – a guarantee in this traditional sense of the word is an obligation by the surety only to pay for the actual damage suffered by the holder of the bond due to the failure of the contractor (or sub-contractor) to perform the work. As it is only a liability to pay for *actual damage*, questions of set-off and counterclaim can be raised in defence against an action on the bond.

Clearly, as a result of this case, a new distinction between the word 'guarantee' and the word 'bond' has arisen, and this will have to be taken account of by those drafting such documents. However, even if the document is described as a bond, and does not contain the word 'guarantee', it may still be held to be in the nature of a guarantee, because it is conditional. Such an interpretation will mean that defences can be raised against any demand for payment. In some cases such a bond will be acceptable to the parties, but it may not be acceptable to purchasers or main contractors who are looking to the bond to provide an immediate source of funds to complete work which remains uncompleted due to default by the party responsible for carrying out the work. The one certainty that emerges from the Trafalgar House case is that there will now be a major upheaval in the evaluation and drafting of performance bonds.

Insurance and engineering contracts

Engineering contracts, as well as many other types of commercial contracts, require several different forms of insurance to be taken care of by the parties. Arrangements may be made quite separately from the contract itself, if the parties so wish. Alternatively, the engineering contract may contain within its 'terms and conditions' conditions about insurance. These will usually describe the nature of the insurance requirements. The different contingencies to be insured against will be described, and the contract will state which party is to carry out the insurance (and pay the premiums). It could be, but need not necessarily be, the same party, who has the burden of effecting all the forms of insurance described, such as insurance of goods in transit or in storage, insurance of works being carried out on site, insurance against injury to employees, insurance against liability to third parties (whether for death or personal injury or damage to property), insurance of the premises of the employer, and professional indemnity insurance.

The value of writing clear conditions about these matters into the contract is that there will be no danger of the parties falling between two stools, as may well occur if each party believes that the necessary insurance will be effected by the other, or that the insurance, in fact, effected by the other is adequate to cover all contingencies. In the case of *AMF International Ltd* v. *Magnet Bowling Ltd* (1968), such a situation occurred, because, as the judge of the High Court stated, the parties were not aware of the legal implications of *subrogation*. Subrogation is a little-known legal principle which works in such a way that if a person, in this case AMF International Ltd, owns equipment or materials on a site where a contractor is carrying out work, it is not enough for the owner of the equipment alone to have insured the equipment. The damage that occurs may be caused by the fault or negligence of another party, such as the contractor. If, as was the case

Negotiating legal and financial matters

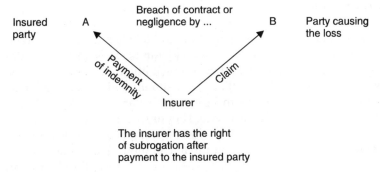

Figure 10.2 *Subrogation*

here, the materials are of high value, the insurer may pay the owner, and then *claim back* the sum paid out from the party responsible for the damage. The law of subrogation is illustrated in Figure 10.2. It is fairly common, in engineering contracts, to find that the parties have made arrangements for the avoidance of subrogation. This can be done by taking out joint insurance, which covers the interests and liabilities of both parties, or it can be done by taking out a policy of insurance which incorporates a waiver of subrogation against the other party, or both methods can be used.

Other matters which may be mentioned in a clause in an engineering contract which deals with insurance include the requirement that the insurance be taken out with an insurance company approved by the employer, and the requirement that the contractor should make proof of the insurance, and of payment of premiums, available for inspection by the employer at reasonable times. These requirements may seem at first sight to be unnecessarily legalistic, but the experiences of some companies have shown the value of such precautions: insurers have been known to become insolvent; contractors have been known to fail to keep up premiums. Finally, there is the question of the sums for which the insurance is to be taken out. These sums may be stated as minimum requirements in the terms of the engineering contract. What the sums should be (for example, public liability may be insured against for as little as £1 million, or as much as £100 million) is a matter for the parties to agree upon. It should be borne in mind by a contractor, that if an indemnity is given by the contractor to the employer in respect of a particular kind of liability, then the sum insured should be at least as much as the maximum amount of the indemnity. If the sum insured is less than this, the full indemnity will still be payable by the contractor, but the uninsured balance will have to be found from the contractor's own funds.

Knock-for-knock clauses

Knock-for-knock clauses are sometimes used in engineering contracts, and relate to liabilities, insurance and indemnities. As they are extremely complicated, and have given rise to much litigation in recent years, they will only be described in outline. They should under no circumstances be included in contracts unless those using them are acting upon the most up to date and legally qualified advice. They are most likely to be used in those engineering contracts which involve a number of parties all working together on site, possibly within a confined space, and possibly in hazardous working conditions. In such circumstances the chances are that it will be difficult to allocate or to apportion blame if an accident occurs. Further, if an accident occurs, it may result in death or personal injury to a number of people, and a potentially large claim or number of claims. The aim of a knock-for-knock clause is to avoid the cost of litigation. If the clause fails to avoid litigation, then it has failed in its aim, and this is what occurred in the case of *EE Caledonia Ltd v. Orbit Valve plc* (1995). This case arose out of the well-known and much publicized accident which destroyed the Piper Alpha oil drilling platform. EE Caledonia Ltd was the employer, formerly known as Occidental Petroleum (Caledonia) Ltd, and had settled a claim for the death of a service engineer provided by Orbit Valve plc. If the clause amounting to a knock-for-knock agreement had been effective, the position would have been that each party would have been liable to bear the loss caused by the death of any of its own employees (this liability would presumably have fallen upon the parties' insurers). The deceased in this case was an employee of Orbit Valve plc, so if the knock-for-knock clause had done the job it was intended to do, Orbit Valve plc would have been liable to repay to EE Caledonia Ltd the damages for which settlement of the claim had been made. *However, the clause was held by the High Court and by the Court of Appeal to be ineffective.*

To understand this case, one has first of all to bear in mind that it would never have arisen if the parties had used the same insurer, with joint insurance against the particular risk, and a waiver of subrogation. If this had been done, the insurance company would have been responsible for the settlement of all such claims, and it would have been immaterial which party actually caused the death, or whose employee the deceased was: but joint insurance and waiver of subrogation is not always possible or seen as practicable by the parties. The knock-for-knock agreement was seen as alternative way of approaching the issue, by making fault irrelevant, and by referring only to the question of *whose employee* was injured or killed.

It will be appreciated by the reader that such an agreement may at first sight appear unjust, because one party may have to pay for the negligence of the other, on the basis of whose employee happens to be the person killed or injured. The theory is, however, that the risk is fairly spread, and the savings in terms of litigation make such clauses worthwhile. The legal problem is that there appears to be no way of writing such clauses other than as indemnities, and indemnities are subject to a long history of precedents which require the draftsman to provide for every contingency and to make it clear above all that the clause is intended to operate *even if one party has to pay or repay money to the other in circumstances brought about by the other's negligence*. A clause which fails to do this will, in many instances, be worthless. It cannot be over-emphasized that the utmost skill and precision in drafting is needed if these clauses are to be used. The problem in the Piper Alpha case was not that such clauses are not permitted, but only that they must be correctly drafted in order to be valid.

Arbitration clauses, and the duration of liability

There is no requirement at all that a commercial contract should contain an arbitration clause of any kind. Traditionally, many such contracts have contained arbitration clauses, and most of the forms published by the engineering institutions have such clauses. However, the vast majority of engineering contracts are made according to private conditions of contract, and many of these do not contain arbitration clauses. The parties to contracts without arbitration clauses have presumably formed the view that legal and other differences or disputes between themselves can best be resolved by settlement out of court, or by litigation, where settlement is not possible. Settlement has obvious recommendations, since the costs are mainly internal and administrative. There need be no publicity at all (except where, after settlement with one party, an indemnity from another is sought, as in the Piper Alpha case). In most cases, after settlement, the two parties are capable of continuing to work or trade together.

Litigation, which can be costly (and which takes place in open court, with the possibility of the case being reported), at first glance seems to have far less to recommend it. Its attraction is that it can be a highly effective way of getting debts settled or paid, and the obverse side of the publicity is that a decision of the High Court, Court of Appeal or House of Lords on a point of law forms a precedent which may give valuable guidance to industry and commerce. In many cases it can be said that on a recurring legal issue, litigation is a form of investment in obtaining a reliable answer to a problem. This sanguine point of view can, however, be taken too far, since the courts are anxious to limit their availability to the public: any case brought before the courts must always be based on real existing facts presenting a genuine triable issue. In the recent case of

Wyco Group plc and Others v. *Cooper Roller Bearings Co. Ltd* (1995), the Chancery Division of the High Court held that the court could not make a declaration of the law upon a theoretical question. The case involved intellectual property rights, and Wyco Group plc and two other companies wished to know whether or not, if they were to manufacture split roller bearings and other associated products similar to those made by Cooper Roller Bearings Co Ltd, Cooper would have a claim based on copyright against them. What was sought, in effect, was a declaration of future non-liability, in circumstances in which the ambit of the law was imperfectly understood. However, this is not something that the courts are prepared to do. This policy of the courts has been in existence for many years and can be justified on the grounds of resources. However, in this particular case, it should be said that the legal issue in question was one which was created by the judges – in the case of *British Leyland Motor Corp. Ltd* v. *Armstrong Patents Co. Ltd* (1986), a case permitting the making of replacement parts in certain circumstances, even though this may infringe the copyright of another party. (Perhaps it was not unreasonable for Wyco Group plc and its fellow plaintiffs to have asked the courts to undo the ambiguity which the courts themselves had brought about, even if the circumstances were hypothetical.)

Where litigation in court does not appear to be worthwhile to the parties, they will usually resolve any differences that may arise between them, either through a negotiated settlement, or by means of arbitration. Arbitration is the referral of an issue or dispute or difference between the parties to an arbitrator. The arbitrator is not a judge of the courts, but is a person who will be selected by the parties themselves. In certain cases the parties will agree each to select an arbitrator, who will then select an umpire, but this latter system, although guaranteeing a certain degree of expertise and impartiality, is likely to be costly. What an arbitration clause, if properly drafted, should always provide for is the possibility that the two parties to the agreement may not be able to agree upon the choice of an arbitrator. The arbitration clause should therefore have a fall-back position under which (in such circumstances) the parties agree to accept arbitration by a person nominated by a third party. Most of the engineering institutions are willing to provide such a service, but this point should be checked before the arbitration clause is drawn up. The aim in providing for the selection of the arbitrator, and in providing for the venue, is to strike a balance of convenience and saving of costs, as well as to obtain expertise in one's particular field. Arbitration has the advantage of privacy, and the rules are rather less formal than those of a court. It is governed, in the United Kingdom, by the Arbitration Act 1950, which provides that the decision of an arbitrator is binding on the parties. The Act also provides for certain procedures, so as to ensure fair play in the conduct of the arbitration.

Venue and choice of law

It is open to the parties to provide in a commercial contract for a particular system of law to apply, in the interpretation and enforcement of the contract; it is also open to them to provide for issues to fall within the jurisdiction of the courts of a chosen country, or for arbitration to be held in a particular place. Thus, a clause may state that the contract is to be governed by and construed in accordance with English Law, and subject to the jurisdiction of the English Courts, or it may provide for arbitration, and for a different system of law to apply. The choices made by the parties must, however, be bona fide, and must have some commercial justification. Arbitration clauses and choice of law clauses are recognized and given effect by the courts. They do not, however, necessarily govern all legal issues arising out of a contract, and some matters, such as injuries to third parties, will remain subject to mandatory rules of law and jurisdiction, no matter what the contract may say. If the parties to a contract do not provide in it for a choice of law and arbitration or the jurisdiction of the courts, then these matters will have to be settled when problems arise, by reference to established principles of law, or, if the parties are in the European Community, by reference to principles set out by treaty under European Community law.

Appeals

Decisions of courts of law are subject to rights of appeal, as will be apparent from the fact that many of the cases reported or discussed in this book are cases which have gone to appeal in the Court of Appeal or the House of Lords. One of the distinctions between arbitration and litigation lies in the question of whether or not an appeal exists. Many arbitration clauses state that the decision of an arbitrator will be final and binding and without appeal. This is particularly true in international arbitrations, and such clauses are valid and effective. The aim is to provide for speed and a saving in costs, by obtaining finality. However, English Law has taken a different view of things, and has come down in favour of appeals, on the ground that mistakes in law by arbitrators should not go uncorrected. The Arbitration Act 1979 therefore provides that all decisions in United Kingdom arbitrations are subject to appeal on points of law. Factual decisions, as such, cannot be appealed against, so a finding by an arbitrator as to whether or not work has been completed, for example, is not appealable: however, a mistake on a point of law, or on a point of interpretation of a contract, would be appealable.

The duration of liability

The issue of the duration of liability is often known to lawyers as the question of 'limitation', but it is not to be confused with other limitations of liability. It is more a question of the legal time limits within which an issue or dispute or action must be pursued, if it is to remain a valid issue or dispute or action. In English law, the governing statute is the Limitation Act 1980 (sometimes referred to as the statute of limitations). This area of the law is based upon the principle that causes of action ought not to last indefinitely, but should, in the interests of certainty of liability, expire after a stated time. To some extent it is possible for parties to a commercial contract to agree upon time limits in the terms of the contract, particularly in the terms of a 'warranty' as to quality: but apart from this, the legal time limits have long been marked out by statute, and the statutes have been changed or consolidated from time to time.

The Limitation Act 1980 operates by reference to different types of causes of action. For the purposes of this book, one need only consider actions arising out of simple contracts, actions arising out of contracts executed as deeds, and actions in the law of tort. The time limits in each case are different. There are two further types of case which also need to be considered separately: these are 'latent damage' caused by negligence, and Product Liability. Under the Limitation Act 1980, different types of proceedings must be 'commenced' within specific periods of time. By 'commenced', it is meant that the official commencement of proceedings, such as taking out a writ or summons, or giving notice of arbitration, must have taken place. So, in practical terms, no matter how many complaints have been made, or how many negotiations as to liability have taken place (unless the writ, etc., has been issued in time), the proceedings will be statute barred, which means that they can no longer be pursued. This may at first sight appear to be a disadvantage, but it does at least provide some element of certainty, and in several of the cases mentioned in this book it will be apparent on closer examination that the party seeking a remedy took full advantage of the statute, by waiting the maximum period of time before deciding to commence proceedings. The categories relevant to this book are as follows:

Actions based upon simple contracts

Actions based upon simple contracts may not be brought after the expiration of six years from the date on which the 'cause of action' accrued – 'simple contract', means any contract other than one executed as a deed under seal, and cause of action means the factual situation which amounts to a breach of contract. So, if a seller fails to deliver goods to a buyer by a contractual date for delivery, the cause of action begins on the due date for delivery: at that moment, the contract is broken.

Actions based on a speciality (that is a contract executed as a deed)

Such actions may not be brought after the expiration of twelve years from the date on which the cause of action accrued. Contracts which are likely to be executed as a deed are loan documents, including mortgages, and bank and parent company guarantees. Bonds may be executed as deeds, but they may also be made as simple contracts.

Debts

Whether a debt arises out of a simple contract, or whether it arises out of a contract executed as a deed, great care is needed in calculating limits of liability. Firstly, it must be remembered that non-payment may be a continuing or repeated breach. This means that one cannot simply count six (or twelve) years from a failure to make payment on a loan: the breach may run right through the period of the loan, until the date for the final payment. It will be from this final date that the period of limitation will begin to run. Further, there is another rule that comes into play, which is that with money debts, written acknowledgement or part payment of the debt, before the expiry of the period of limitation, starts time running afresh. So, if five years have gone by, and payment is outstanding, and if the debtor pays half the debt, or acknowledges the existence of the debt, in writing, this will start time running afresh, for the full period of limitation.

Actions founded on tort

Here we have to distinguish between those cases which involve personal injuries and those which involve damage to property or other material losses. The period for personal injuries is *three* years from the date of the injury (or from the date when its existence becomes apparent). In other cases the period is *six* years, but it differs from the period under the law of contract, since it commences *when the damage occurs*. So, for example, if a seller delivers defective goods to a buyer, the breach of contract, and the running of the period of limitation, will take place at this moment. However, if the defects cause damage, for instance by a fire or explosion, the period of limitation in tort will only start at the moment of the damage. This means that there can be advantages in bringing an action under the law of tort, rather than under the law of contract, although there can also be disadvantages which were alluded to in an earlier chapter.

168 *Negotiating legal and financial matters*

Product Liability

This law is of recent creation, and is common to all systems of law in the European Community. Actions for Product Liability are subject to a period of limitation, which is similar to the period for personal injuries caused by negligence: but there is an overall limit of ten years, dating from the date of supply. So if, for example, an engine is supplied which causes injury to a person who comes into contact with it, that person may bring an action for damages for Product Liability. The action may be brought against the producer or importer into the European Community, or brander of the product. The period for commencement of the action is three years, dating from the date of the injury, or dating from the injured person having knowledge of his injury, but once the engine has reached an age of seven years dating from the date of supply, the ten-year limit starts to bite: the injured party must stay within the ten-year limit, and his three years will be reduced to that extent. So, if the injury were to occur in the ninth year after supply, the injured party would have less than one year in which to claim. However, if the ten-year period has totally expired, all is not lost for the injured party: he cannot claim for Product Liability, but may still bring a claim under the ordinary laws of

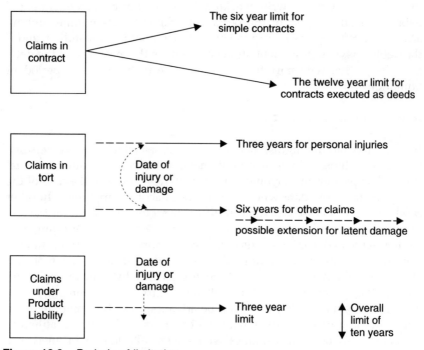

Figure 10.3 *Periods of limitation*

negligence, and avail himself of the period of limitation which applies to torts, and which is three years without the overall ten-year limit. However, the range of possible defendants is more limited with negligence.

The picture which begins to take shape, as to the comparison of different rules under the Limitation Act, can be seen in Figure 10.3.

The Latent Damage Act 1986

One further point needs to be made, which is of some importance in engineering contracts. Several cases in the past few decades have raised the possibility that the ordinary periods of limitation may give rise to injustice. Even allowing for the fact that the six-year period in tort only begins when the damage first occurs, there is the possibility that the damage may occur without the party suffering it being aware of it – for example, cracks may occur in inaccessible parts of works such as foundations or high chimneys, metal fatigue or hidden corrosion may occur. At a certain moment this may cause the building or structure or plant or works to be 'damaged' (although the precise meaning of that word is not entirely free from doubt), but the customer may not know about the damage. Such a case did in fact happen in *Pirelli General Cable Works Ltd* v. *Oscar Faber and Partners* (1982). The single most important point in this case was the fact that it perfectly illustrated the problem of the Limitation Act 1980. A chimney had been designed, which was 48.8 metres high, and which began to crack internally when put to use, due to unsuitable materials having been used for its refractory lining. The cracks were not discovered until eight years after commissioning, so that it was not possible to sue the main contractor. The 'breach of contract' would have taken place on commissioning or handing over, and the six-year rule then applied. The action which was commenced in tort, under the law of negligence, was against the consulting engineers. No liability was proved, since the defence of the consulting engineers was that the action was out of time. The dilemma for the courts was whether the date from which the six-year rule in tort was to run was from when the damage 'occurred', or from when it was first discovered. In the Pirelli case the two dates were very different, since, with scientific backdating it could be shown that the damage occurred in 1970, while the discovery was in 1977, and the writ was issued in 1978. The House of Lords held that the correct date for commencement of limitation was the date when the damage *occurred*. This was seven years and eleven months before the issue of the writ, so the writ was out of time.

In 1986, largely to remedy the perceived injustice of this case, the Latent Damage Act 1986 was passed. This Act applies to latent damage caused by negligence (other than personal injuries). It allows the claimant a further three years from when the latent damage is discovered or from when the claimant had knowledge of the material facts about the damage. The Act applies to any qualifying case which was not already statute-barred before it came into force in 1986. There is an overall limit under this Act of fifteen years dating from the date of the action which is alleged to be negligent; so even with latent damage time limits do not continue indefinitely.

A legal and practical question: when should inspections take place?

It has already been demonstrated that legal time limits for action are an intricate matter, and where action is contemplated qualified advice is needed in each particular case. However, those involved in commerce and in engineering projects may wonder if, on the strength of this discussion, there is any practical guidance which emerges that can be put to everyday use. The answer lies in the nature of *inspections* which the prudent customer would carry out at various periods after delivery or handing over of works or equipment. Where practicable, such inspections should permit the purchaser a breathing space so as to be able to decide upon the necessary course of action. Obviously, an inspection at the end of a warranty or defects liability period is advisable. Beyond this, the six-year rule which applies to most contracts would seem to indicate that an inspection some months before the period has expired would have possible benefits, since any residual statutory rights which the purchaser may have could then be enforced. With works which might harbour latent damage, inspections at three-yearly intervals, up to the fifteenth year, would be a way in which to take advantage of the extension of the normal time limits which is given by the Latent Damage Act 1986. It must be emphasized, however, that neither a purchaser's statutory rights, nor the Latent Damage Act 1986, give the purchaser the right to claim in respect of defects which are due solely to age, wear and tear, or poor maintenance. The time limits are about the final dates on which action in respect of breach of contract, or a tort such as negligence, can be taken. The mere fact that plant or equipment or buildings are showing defects or deterioration after a period of time, does not of itself mean that there has been a breach of contract or any form of negligence. Evidence of breach of contract, or of negligence, would have to be brought by the party claiming: this would have to show, for example, a defective design or incorrect choice of materials, or workmanship that could not have withstood the expected use for the expected period of time the item was to be in service. In the case of *latent damage*, this is not quite the same thing as *latent defects*: the 1986 Act requires that *damage* should have been caused by *negligence*, before the extension of time comes into play. Exactly where the line is drawn between a defect and damage is for the judge to decide on the facts of each case. In the case of the chimney in the *Pirelli* case, for example, one could say that the lining was defective as installed, but that the damage did not occur until it cracked some months later. However, having given this example, no pretence will be made that it is easy to distinguish between the two concepts.

The future

Standardization or diversity?

One of the questions which has existed for several decades, and which industry has yet to resolve, is whether engineering contracts should be made according to standard, national forms of contract, such as those published by the well-known engineering institutions, or whether it is more beneficial for each party to put forward its own private terms of contract and use these as a starting point for negotiations. In favour of the latter, it may be said that each party can develop its own internal policies, and the contracts that subsequently are made will reflect those policies to a greater or lesser extent, according to the outcome of the negotiations. This way of making contracts is certainly the one most likely to be adopted in major, and particularly in international, contracts where the purchaser will aim to use its purchasing power to obtain the most favourable bargain available. Conversely, it can also be said that in-house conditions of contract are most likely to be used in the smaller sales and engineering contracts. This is largely for reasons of cost and convenience. Each party, rather than entering into discussions about institutional forms of contract, will find it simpler and more cost-effective (in smaller contracts) to issue quotations, purchase orders and confirmation documents on forms which contain their own conditions of contract. This often, inevitably, leads to the battle of the forms, and to uncertainty as to what end result has been reached. Yet industry has shown a surprising ability to live with these problems.

Institutional forms of contract, such as the excellent Model Forms 1, 2 and 3, issued by the Institution of Mechanical Engineers, Institution of Electrical Engineers, and Association of Consulting Engineers, as well as the long established JCT forms of the Joint Contracts Tribunal, have much to recommend them. They are intended to be neutral, as far as this is possible; they are kept up to date and reviewed from time to time; they contain prepared forms for the use of the parties, such as Tender forms, and forms of certificate; and, so long as both parties make it clear from the start that their use is contemplated, they avoid the battle of the forms. If their use is not as widespread as might have been thought, this might, it is suggested, be due to a reluctance of the engineering industry to standardize itself in the way in which this has occurred in the building industry. Even within the building industry, standardization is imperfect, but there is at least a measure of acceptance that JCT forms are to be expected as the most likely way of forming such contracts. At present no such consensus exists in the engineering industries or among their clients, perhaps because no one organization or body has taken on the responsibility of bringing it about.

Plain English

A glance at almost any section of this book will reveal that almost any problem arising in a commercial contract, no matter how simply set out, will sooner or later throw up abstract legal issues which will need complex ideas and analysis and language to reach a conclusion. A relatively simple matter, such as that of delay in delivering goods or completing work, can raise issues such as whether or not time is 'of the essence', issues about liquidated damages and their distinction from penalties, as well as issues of set-off and counterclaim. The question is whether dealing with these matters is made any easier if the contract is in 'plain English'. At first sight, plain English has much to recommend it, particularly when it is borne in mind that an engineering contract has to be read by, and used by, managers who are not legally qualified. At the beginning of this book, it was stated that a commercial contract is intended to be a document that facilitates commercial planning, and it should help the parties to achieve mutual understanding and expectations. This statement holds good, and those parts of an engineering contract which are intended for the information of the parties who have to work with it should, as far as possible, be in the plainest possible language. This sentiment is entirely in keeping with the spirit of the EC Directive on Unfair Terms in Consumer Contracts, in which Article 5 requires all terms of contracts made with consumers to be in 'plain intelligible language'. This has no binding effect where the customer is a commercial person or body as opposed to a private consumer, but the Directive is likely to set a trend towards simpler language in contracts.

Nevertheless, plain language brings its own problems. Complex legal expressions were not invented as an obscure code to baffle the uninitiated, but instead grew up as the most precise way in which complicated propositions could be stated, preferably so that both the person drafting them, and the judge interpreting them would find the same meaning in them. To the unqualified reader, the expression 'waiver of subrogation' may be so obscure as to be meaningless; yet to the expert its meaning is so precise that the same sense could not be conveyed by any alternative words. The same is almost certainly true of many expressions pertaining to such matters as insolvency, intellectual property, bailment, and trusts and fiduciary relationships. Plain language is desirable, but it may lead to loss of precision. It is problems of this kind that those whose task it is to draft contracts in plain English must confront.

Will commercial contracts become more cooperative and less adversarial?

In answer to this question, it has to be said that if parties wish to approach business transactions in a less adversarial manner, and to see themselves as having interests in common – rather than as parties with opposing interests – it is not simply the contract and its terms that must be looked at, but the entire commercial background to the transaction and the relationship. Certainly it is true that a contract with particularly adversarial terms and an unnecessarily legalistic approach can hinder, rather than help, development of cooperation between parties to a commercial venture, but that is by no means the whole story. Cooperation grows best where businesses see themselves as long-term trading partners: where they see each venture or project as a collaborative effort; where there is a system of accredited suppliers and recognition is given for the ability and willingness of parties to work with one another; where solutions to problems are sought jointly; and where training or staff development is carried out across the boundaries of contractor and purchaser. These are deep and far-reaching issues which lie beyond the scope of this book, and which a work of this kind can only hope to allude to in passing.

Forms of contract, and the terms and conditions contained in them can make their contribution towards a different approach in business relationships. The problem is that the development of such forms of contract is not very far advanced at present. It is for reasons of this kind, no doubt, that the Department of the Environment and the construction industry funded a study by Sir Michael Latham of the construction procurement process in the United Kingdom. The report resulting from this study, called 'Constructing the Team', published in 1994, has given rise to several consultation papers from the Department of the Environment, one of which is titled 'Fair Construction Contracts', and which was issued in May 1995. This paper is primarily concerned with the construction industry, as compared with engineering in the wider sense, but many of the ideas contained in it have potential application to all commercial contracts of the kind with which the manufacturing industry is concerned. Of particular interest is the fact that the paper notes the prevalence of non-standard terms being used to amend standard forms of contract: because of this there are suggestions in the paper that it may have to be made impossible for parties to standard form contracts to amend or to delete certain essential provisions. For this to work in practice there would have to be officially recognized forms of contract, and legislation to make certain terms mandatory. 'Essential terms', as identified in the consultation paper, include terms about dispute resolution, the right of set-off, prompt payment (the issue of interest for overdue payment may arise again here), and protection against insolvency. Possibly other terms may be classified as 'essential' in this sense. It should be emphasized that this review is, at the time of writing, still continuing, and there is no certainty as to when, if at all, the Government might be prepared to take what would be one of the most radical steps ever taken to limit freedom of contract.

Appendix 1

Designing and structuring an engineering contract

The form and structure of an engineering contract will always depend upon the size and complexity of the project, which will, in turn, dictate the number of documents required. At its simplest this will require no more than an order and an acceptance, with a number of conditions attached. In this appendix we are looking at the kind of project which will need some or all of the following documents:

Invitation to Tender
Tender
General Conditions of Contract
Special Conditions of Contract
Specification
Drawings
Other technical documents
Form or Memorandum of Agreement
Specimen Form of Bond or Guarantee

Identification of the contract documents

It is essential that all the documents which are intended to become part of the contract should be identified. This should be done in the form or memorandum of agreement, and in the conditions of contract. It is not essential, but may be thought to be desirable, to include in the conditions of contract a clause which states an order of priority of documents, in case there should be any conflict between any provisions of any documents forming part of the contract. It is normal to provide that the memorandum of agreement (or equivalent document) should be the document having priority over all others, followed by the special conditions, the general conditions, the specification, and other specified documents, in that order.

Definitions

Good conditions of contract should attempt to define matters which might otherwise be difficult to interpret, such as:

    'acceptance'
    'authorized person'
    'change'
    'contract'
    'contract price'
    'contract value'
    'engineer'
    'firm price'
    'fixed price'
    'month'
    'specification'
    'sub-contractor'
    'supplier'
    'week'
    'works'

The conditions

Standard, or general conditions of an engineering contract may set out matters of the kind that appear in the list that follows. No particular order of setting out conditions of contract has become uniformly accepted, but an order which follows some form of logical time sequence or thought process is desirable. In the list given, the order takes the pattern that follows:

    Obligations
    Rights
    Contingencies
    Completion and/or termination
    Settlement of disputes

A full list of conditions set out in this order would include the following (where further details are required, they would be set out in Special Conditions):

    Definitions
    The parties
    Contractor's duties
    Employer's duties

The engineer
Assignment and sub-contracting
The site: inspection, access, security
Date for commencement
Date for completion
Basis of price
Terms of payment
Performance bond or guarantee
Submission and approval of programme
Submission and approval of drawings
Health and safety obligations
Tests, acceptance and certificates
Secrecy
Intellectual property rights
Ownership of goods and materials
Warranties, defects liability, limits of liability
Variations or changes
Progress, delay, liquidated damages
Extension of time, *force majeure*
Claims for additional payment
Insurance
Allocation of risks, indemnities
Suspension of work
Employer's rights to terminate the contract
Contractor's rights to terminate the contract
Arbitration/jurisdiction of the courts
Law applicable to the contract

Appendix 2

Statutes and other legislation

(in chronological order)
- 1677 Statute of Frauds
- 1950 Arbitration Act
- 1967 Misrepresentation Act
- 1974 Consumer Credit Act
- 1977 Patents Act
- 1977 Unfair Contract Terms Act
- 1979 Sale of Goods Act
- 1979 Arbitration Act
- 1980 Limitation Act
- 1982 Supply of Goods and Services Act
- 1985 EC Directive 1985, No. 85/374 EEC on the approximation of the laws, regulations and administrative provisions of the member states concerning liability for defective products
- 1986 Insolvency Act[1]
- 1986 Latent Damage Act
- 1987 Consumer Protection Act
- 1988 Copyright, Designs and Patents Act
- 1989 Law of Property (Miscellaneous Provisions) Act
- 1990 Contracts (Applicable Law) Act
- 1993 EC Directive on Unfair Terms in Consumer Contracts
- 1994 Sale and Supply of Goods Act
- 1995 Unfair Terms in Consumer Contracts Regulations
- 1995 Sale of Goods Amendment Act

Appendix 3

Cases

Note: A number of useful cases on commercial and engineering contracts are unreported. This is mentioned where applicable in the table of cases that follows. Where cases have appeared in *The Times*, or in other newspapers, or in *Current Law*, this has been stated. Otherwise the law reports in which individual cases are reported appear under the following abbreviations:

AC	*Appeal Cases*
ALJR	*Australian Law Journal Reports*
All ER	*All England Law Reports*
BLR	*Building Law Reports*
Ch	*Law Reports, Chancery Division*
DLR	*Dominion Law Reports*
EG	*Estates Gazette*
Lloyd's Rep	*Lloyd's Law Reports*
NLJ	*New Law Journal*
NZLR	*New Zealand Law Reports*
P&CR	*Property and Compensation Reports*
QB	*Law Reports, Queen's Bench Division*
Tr LR	*Trading Law Reports*
WLR	*Weekly Law Reports*

AMF International Ltd *v.* Magnet Bowling Ltd (1968) 2 All ER 789

Armour *v.* Thyssen Edelstahlwerke AG (1990) 3 All ER 481

Aswan Engineering Establishment Co. *v.* Lupdine Ltd (Thurgar Bolle Ltd, third party) (1987) 1 All ER 135

Axel Johnson Petroleum AB *v.* M. G. Mineral Group AG (1992) 1 WLR 270

Beoco Ltd v. Alfa Laval Co. Ltd and Another (1994) 4 All ER 464
Bevan Investments Ltd v. Blackhall and Struthers (1973) 2 NZLR 45
Blackpool & Fylde Aero Club v. Blackpool Borough Council (1990), *The Independent*, June 12
Bolton v. Mahadeva (1972) 1 WLR 1009
Bond Worth Ltd, Re (1979) 3 WLR 629, (1980) Ch 228
Bramall and Ogden Ltd v. Sheffield City Council (1983) Unreported
Brinkibon Ltd v. Stahag Stahl und Stahlwarenhandelsgesellschaft MbH (1983) AC 34
British Leylard Motor Corp. Ltd v. Armstrong Patents Co. Ltd (1986) (1986) 2 WLR 400.
British Steel Corp. v. Cleveland Bridge and Engineering Co. Ltd (1984) 1 All ER 504
Butler Machine Tool Co. Ltd v. Ex-Cell-O Ltd (1979) 1 All ER 965
Cable Ltd v. Hutcherson Ltd (1969) 43 ALJR 321
Cammell Laird and Co. Ltd v. Manganese Bronze and Brass Co. Ltd (1934) AC 402
Cana Construction Co. Ltd v. The Queen (1973) 37 DLR 418
Circle Freight International Ltd v. Medeast Gulf Exports (1988) 2 Lloyd's Rep 427
Courtney and Fairbairn Ltd v. Tolaini Bros (Hotels) Ltd (1975) 1 All ER 716, 1 WLR 297
Croudace Construction Ltd v. Cawoods Concrete Products Ltd (1978) 2 Lloyd's Rep 55
Dawber Williamson Roofing Ltd v. Humberside County Council (1979) *Current Law* 212
Dominion Mosaics Tile Co. Ltd v. Trafalgar Trucking Co. Ltd (1989) *The Times*, March 2
Edward Owen Engineering Ltd v. Barclays Bank International Ltd and Umma Bank (1978) 1 All ER 976, 2 Lloyd's Rep 166
EE Caledonia Ltd v. Orbit Valve plc (1995) 1 All ER 174
Esmil Ltd v. Fairclough Civil Engineering Ltd (1981) 19 BLR 129
Evans and Son (Portsmouth) Ltd, J. v. Andrea Merzario Ltd (1976) 2 All ER 930
Grayston Plant Ltd v. Plean Precast Ltd (1976) Unreported
Hanson W. (Harrow) Ltd v. Rapid Civil Engineering Ltd (1987) 38 BLR 106
Harbutt's Plasticine Ltd v. Wayne Tank and Pump Co. Ltd (1970) 1 QB 447, 1 All ER 225
Harvey v. Ventilorenfabrik Oelde (1989) 8 Tr LR 138
Helstan Securities v. Hertfordshire County Council (1978) 3 All ER 262
Highway Foods International Ltd, Re (1994) *The Times*, November 1
Hong Kong and Shanghai Banking Corp. v. Kloeckner & Co. AG (1989) 3 All ER 513, (1990) 2 QB 514

IBA v EMI Ltd, IBA *v.* BICC Ltd (1980) 14 BLR 1
Ibmac Ltd *v.* Marshall Ltd (1968) 208 EG 851
Interfoto Picture Library Ltd v Stiletto Visual Programmes Ltd (1988) 1 All ER 348
Intertradex SA *v.* Lesieur-Tourteaux SARL (1978) 2 Lloyd's Rep 146
Junior Books Ltd *v.* The Veitchi Co. Ltd (1982) 3 All ER 201
Kleinwort Benson Ltd *v.* Malaysian Mining Corp. Berhad (1989) 1 WLR 379
Kolfer Plant Hire Ltd *v.* Tilbury Plant Ltd (1977) *The Times*, May 18
Koufas *v.* Czarnikow (1967) 3 All ER 686
Linden Garden Trust Ltd *v.* Lenesta Sludge Disposals Ltd (1994) 1 AC 85
Lombard North Central plc *v.* Butterworth (1987) 2 WLR 7
Lubenham Fidelities and Investment Co. Ltd *v.* South Pembrokeshire District Council and Wigley Fox (1986) *The Times*, April 8
McDougall *v.* Aeromarine of Emsworth Ltd (1958) 3 All ER 431
McGrath *v.* Shah (1989) 57 P&CR 452
MacJordan Construction Ltd *v.* Brookmount Erostin Ltd (1991) *The Times*, October 29
Mitchell (George) (Chesterhall) Ltd *v.* Finney Lock Seeds Ltd (1983) AC 803
Muirhead *v.* Industrial Tank Specialities Ltd and Others (1985) 3 All ER 705
Orion Insurance Co. *v.* Sphere Drake Insurance (1990) *The Independent*, February 1
Pacific Associates Inc. and R. B. Construction Ltd *v.* Baxter and Others (1988) 1 QB 993
Parkinson & Co. Ltd *v.* Commissioners of Works (1950) 1 All ER 208
Perini Pacific Ltd *v.* Greater Vancouver Sewerage and Drainage District (1966) 57 DLR 307
Photo Production Ltd *v.* Securicor Ltd (1980) AC 827
Pirelli General Cable Works Ltd *v.* Oscar Faber and Partners (1983) 2 AC 1
Pyrene Co. Ltd *v.* Scindia Steam Navigation Co. Ltd (1954) 1 Lloyd's Rep 321
Redler Grain Silos Ltd *v.* BICC Ltd (1982) 1 Lloyd's Rep 435
Rees Hough Ltd *v.* Redland Reinforced Plastics Ltd (1984) *New Law Journal*, August 17 1984
Roberts & Co. Ltd *v.* Leicestershire County Council (1961) Ch 555
Ruxley Electronics and Construction Co. Ltd *v.* Forsyth (1995) 3 WLR 118
Simaan General Contracting Co. *v.* Pilkington Glass Ltd (1988) 2 WLR 761
Smith *v.* South Wales Switchgear Ltd (1978) 1 All ER 18

St Albans City and District Council *v.* International Computers Ltd (1994) *The Times,* November 11

St Martins Property Corp. Ltd *v.* Sir Robert McAlpine Ltd (1994) 1 AC 85.

Stewart Gill Ltd *v.* Horatio Meyer Ltd (1992) 2 All ER 257

Tate & Lyle Food and Distribution Ltd v Greater London Council (1982) 1 WLR 149

Teheran-Europe and Co. Ltd *v.* S. T. Belton (Tractors) Ltd (1968) 2 All ER 886

Trafalgar House Construction (Regions) Ltd *v.* General Surety & Guarantee Co. Ltd (1995) 3 WLR 204

Trollope and Colls and Holland and Hannen and Cubitts Ltd *v.* Atomic Power Constructions Ltd (1962) 3 All ER 1035

Walters *v.* Whessoe Ltd and Shell Refining Co. Ltd (1960). Reported in note to AMF Ltd v Magnet Bowling Ltd (1968) 2 All ER 816

Widnes Foundry Ltd *v.* Cellulose Acetate Silk Co. Ltd (1933) AC 20

Williams *v.* Roffey Bros & Nicholls (Contractors) Ltd (1990) 1 All ER 512

Wyco Group plc and Others *v.* Cooper Roller Bearings Co. Ltd (1995) *The Times,* December 5

Young & Marten Ltd *v.* McManus Childs Ltd (1969) 1 AC 454

Appendix 4

List of engineering institutions and their contracts

There are several institutions which have produced forms of engineering or construction contracts. Some of these have been mentioned in the text, (although failure to mention an institution is in no way a reflection upon that institution or its importance). Where they have been mentioned, and in particular where an abbrieviated reference has been made, the following explanatory note may be found helpful.

The Association of Consulting Engineers
Alliance House, 12 Caxton Street, London SW1H 0QL
   The Association Produces the *ACE*. Conditions for the engagement of a consulting engineer. It is one of the publishers of the *ICE Conditions of Contract*, and also one of the recommending bodies for the Model Forms of General Conditions of Contract.

The Institution of Civil Engineers
Great George Street, London SW1
   The Institution sponsors, approves and publishes the *ICE Conditions of Contract*.

The Institution of Chemical Engineers
George E. Davis Building, 165/171 Railway Terrace, Rugby CV21 3HQ
   The Institution produces and publishes Model Forms of Conditions of Contract for Process Plants.

The Institution of Electrical Engineers
2 Savoy Place, London WC2
   The Institution, together with the Institution of Mechanical Engineers, has formed the Joint I Mech E/IEE Committee on Model Forms of

Engineering Contract, and has published the Model Forms of General Conditions of Contract, Known as *MF/1*, *MF/2*, and *MF/3*. The numbering system relates to the detail of the forms of contract and their suitability for particular types of project.

The Joint Contracts Tribunal

This body has a number of constituent bodies, including the Royal Institute of British Architects and Association of Consulting Engineers and Confederation of Associations of Specialist Engineering Contractors. It produces the forms of Building Contract known as *JCT*. These are published by RIBA Publications Ltd. There are numerous forms of JCT contract, depending upon the nature of the works and the style of contract required, but the details as to suitability for any particular purpose are better dealt with in a work on building contracts. They are mentioned in this work because some of the principles of commercial and engineering law set out in this book have been formulated by the courts in the context of JCT contracts.

Glossary

The trend is towards writing contracts, including engineering contracts, in plain, intelligible language. However, it is not easy to keep commercial contracts or their interpretation or explanation entirely free from traditional legal expressions. The glossary that follows may be found useful in understanding the meaning of terms commonly encountered in engineering contracts.

Acceptance   This expression takes its meaning from its context. It may mean the acceptance of an offer in such a way as to form a contract, or it may mean something which has only commercial significance, but which falls short of being a binding contract, for example, an acceptance which is qualified. It can also refer to things which occur in the later stages of a contract, such as acceptance of goods or acceptance of work after tests have been satisfactorily carried out.

Acknowledgment   This word should be used with greatest care: it may amount to a mere confirmation that something has been received, or it may amount to an acceptance of an offer. It could vary the offer, and so be a counter-offer.

Administrator   An official appointed by a court under an Administration Order made under the Insolvency Act 1986.

Advance Payment Bond   A bond or guarantee given to secure advance payments made by a buyer.

Arbitration   A form of adjudication of disputes or differences between parties to contracts. An arbitration agreement is an agreement to submit disputes or differences or questions arising between parties to a person or persons other than a court of law.

Bailee   One who holds goods on behalf of another. Hirers are bailees. Purchasers who hold goods subject to a retention of title in favour of the seller are bailees.

Bid Bond A bond submitted by a party making a tender at the time of making the tender, usually as a security that the tender will not be withdrawn or amended. Sometimes known as a tender bond.

Change See **Variation**

Charge A form of security created by a company in favour of another person, usually a creditor of the company. The charge confers rights over the charged assets of the company: 'fixed charge', or 'floating charge'.

Condition A term of a contract. In the strict sense of the word, a condition is a major term of a contract, but the word is not always used in its strict sense.

Consequential Loss This expression has no accurate meaning unless it is defined in a contract. However, it is generally understood to mean loss which does not flow directly and naturally from a breach of contract, but which arises as a secondary consequence of a breach of contract.

Determination This word is sometimes used as an alternative to the word 'termination', for example, in 'consequences of determination of the contract by the purchaser'.

Entire Agreement Clause A clause which is sometimes put into a contract to make it clear that the parties may not look for the terms of the contract outside the documents described as constituting the entire agreement.

Entire Contract A contract in which the whole price is allocated to entire performance of the contract, and in which the work must be completely (or at least substantially) performed before any liability to pay arises.

Force Majeure Circumstances beyond the parties' control, as defined and agreed upon in the contract.

Frustration Common law principle that a contract will come to an end if circumstances beyond the control of the parties make performance illegal, impossible, or radically different from what was originally contemplated.

Guarantee This word has two entirely different legal meanings. Its first meaning is an undertaking made by one person to answer for the debt or default of another person, as in 'parent company guarantee', or 'bank guarantee'. Under the Statute of Frauds 1677, this is one of the forms of contract which must be evidenced in writing. The second meaning of the word is similar to that of the word 'warranty' in relation to goods or services.

Indemnity An undertaking made by one person to keep another person free from loss.

Instruction to Proceed A request or instruction or authorization made by one person to another, requiring the delivery of goods or performance of services. Whether or not it amounts to a contract depends upon the context.

Intellectual Property Form of recognized and protected property rights in ideas or the presentation of them; for example, patent and copyright.

Latent Damage Damage not apparent on reasonable inspection of the property. The Latent Damage Act 1986 provides for special time limits for claims.

Latent Defects These are defects in goods or work done under a contract, which are not apparent on reasonable inspection or the carrying out of tests (they do not necessarily cause damage). The contract may make special provision.

Letter of Acceptance An acceptance of a tender. May be unconditional and amount to a contract.

Letter of Comfort Letter issued, for example, by a parent company of a debtor or a contractor, and intended to facilitate a loan to the debtor or the award of a contract to a contractor. Normally falls short of a binding guarantee, but the effect depends upon the precise wording.

Letter of Intent Letter conveying a commercial decision to proceed with a transaction, but usually falling short of an offer or an acceptance.

Lien Legal right, arising in certain circumstances, to retain certain goods which one has in one's possession, even though they may be owned by another person.

Limitation (Period of/statute of) Expression used to define time limit(s) within which a person may bring an action against another.

Liquidated Damages Definite and ascertained sum or scale agreed upon at the time that a contract is formed as damages for certain breaches of the contract.

Liquidator Person appointed to wind up a company.

Lump Sum Contract Contract to complete work for a lump sum, as opposed to a contract under which the price is to be arrived at by subsequent measurement of work and materials.

Mitigate, Duty to Rule relating to damages for breach of contract: the party claiming damages must act reasonably to keep his loss to a minimum.

Penalty Contractual provision attempting to impose penal consequences upon a party who is in breach of contract. Penalties are void under English law.

Performance Bond Bond or guarantee given by a surety, such as a bank or insurance company, as a security for the performance of a contractor's obligations under a contract.

Performance Test(s) One or more tests in an engineering contract, usually performed after the delivery and handing over of the works to the purchaser.

Product Liability Liability for injury or damage caused by a defective product. The expression is used especially with regard to the form of liability which can arise irrespective of any fault or of any contractual relationship. Arises in the EC as a result of the EC Directive of 1985, and in the UK under the Consumer Protection Act 1987.

Quantum Meruit Right to be paid a reasonable sum for work done or goods supplied. May exist where there is no contract, or where a contract is frustrated, or as an alternative remedy in the case of a breach by a purchaser.

Receiver Person appointed to receive assets of a company and to pay secured creditors. In certain cases, is known formally as an Administrative Receiver.

Repudiation of a Contract Indication by a party to a contract of a definite intention no longer to be bound by the contract. May be inferred from conduct: serious breach of a contract.

Rescission of a contract The legal termination of a contract, in accordance with the express or implied terms of the contract. May occur, for example, on grounds of breach of condition, or on grounds of misrepresentation of material facts prior to the making of the contract.

Retention Money Money retained or deducted by the purchaser from sums due to the contractor, as security for the performance of the contract.

Retention of Title Form of security for sellers of goods, arising where there is a term in the contract of sale that, notwithstanding the delivery of the goods, the seller is to remain the owner of the goods until payment by the purchaser, or until some other specified event.

Sub-contractor Normally used to mean a contractor to a main contractor. The expression has a number of possible variants and should be carefully defined.

Subrogation The right of an insurer who pays the insured party to avail himself of any right or remedy which the insured party may have against another person in respect of the loss or damage for which the payment is made.

Substantial Performance   Test for determining what amounts to the complete performance of work under a building or engineering contract. In an 'entire' contract, work is normally considered to have been performed so as to earn payment if it has been substantially performed.

Supplier   This expression depends upon definition and context. May mean a seller of goods, or a supplier under a contract of hire or hire-purchase. Sometimes used to mean a sub-contractor, and sometimes used to mean a supplier in a 'supply only' contract.

Term of a Contract   A contractual undertaking. Terms may be classified as conditions or as warranties, or their classification may be 'innominate', that is, indeterminate.

Tort   Civil injury or wrong of a kind recognized by law, and for which the law provides a remedy. Examples of torts are: defamation, negligence, nuisance, trespass, and breach of statutory duty.

Variation   Expression normally used to mean alteration to or addition to, or omission of, work to be done under a contract. See also **Change**.

Variation of a Contract   This differs from variation, in so far as it concerns alteration of the *terms* of an existing contract, rather than operating within the terms of a contract to vary the work or specification. Variation of a contract can only be done by mutual agreement by the parties to the contract, and to be valid, it must be supported by fresh consideration.

Waiver   Concession by one party to a contract to the other. Intimation by a party to a contract that he will give up a right under the contract or will accept less than full performance. May be made expressly or by conduct. Does not require fresh consideration for its validity. Once a waiver is made it cannot be withdrawn.

Warranty   The word has several possible meanings. Its standard legal meaning is a term of the contract which is less important than a condition; breach of warranty does not entitle the party to whom the warranty is given to terminate the contract. However, the word may be used in a stronger sense, depending upon the context. In *insurance* contracts, warranties are important terms, and the insurer may repudiate liability if warranties are untrue or are not complied with.

Select Bibliography

The list that follows is partly given to indicate specialist works in which more details of particular topics dealt with in this book can be found, and partly to suggest further and alternative reading for those who wish to further pursue the subject of commercial contracts. It is, necessarily, a very restricted list and reflects, to a large extent, the author's personal preference.

Atiyah, P. S. (1995) In *The Sale of Goods* (J. Adams, ed.), Pitman.
Bainbridge, D. (1994) *Intellectual Property*, 2nd edn. Pitman.
Bainbridge, D. (1995) *Cases and Materials in Intellectual Property Law*. Pitman.
Benjamin, J. P. (1992) *Sale of Goods*, 4th edn. Sweet & Maxwell.
Cheshire, G. C., Fifoot, C. H. S. and Furmston, M. P. (1991). *Law of Contract*. Butterworth-Heinemann.
Clark, A. M. (1989) *Product Liability*. Sweet & Maxwell.
Dobson, P. and Schmitthof, C. (1991) *Charlesworth's Business Law*. Sweet & Maxwell.
Goode, R. M. (1983) *Payment Obligations in Commercial and Financial Transactions*. Sweet & Maxwell.
Henderson, S. (1994) *Management for Engineers*. Butterworth-Heinemann
Hudson, A. A. (1995) In *Building and Engineering Contracts*, 11th Edn, (N. Duncan Wallace, ed.), Sweet & Maxwell.
Jacob, R. and Alexander, D. (1993) *A Guide to Intellectual Property*. Sweet & Maxwell.
Kader, A., Hoyle, K. and Whitehead, G. (1996) *Business and Commercial Law*, 'Made Simple Series'. Butterworth-Heinemann.
Markesinis, B. S. and Munday, R. J. C. (1992) *An Outline of the Law of Agency*, 3rd edn. Butterworths.
Painter, A., Lawson, R. and Smith, D. (1992) *Business Law*, 2nd edn. Butterworth-Heinemann.
Treitel, G. H. (1995) *Law of Contract*, 9th edn. Sweet & Maxwell

Index

Acceptance, 185
  of offers, 4, 6, 185
  of goods, 185
  of work, 31, 185
Acknowledgment, 5
  of order, 5, 18
  of tender, 5
Administration order, 185
Administrator, 185
Advance payment, 127, 128
Advance payment bond, see Bond
Agency, 20, 138–40, 146, 147
Approval of drawings, 87
Arbitration, 163–5, 185
Architect, 21
Ascertainment of goods, 128
Assignment, 36
Authority of agents, 20
Authority to vary work, 27, 29

Back-to-back contracts, 143
Bailee, 185
Bailment, see Bailee
Battle of the forms, 8, 10
Bond, 152–60
  advance payments bond, 153
  bid bond, 153
  performance bond, 46, 154
  tender bond, 153
  warranty bond, 154
Breach of contract, 98

Canadian cases, 38, 142
Carriage, contract of, 57–62

Cause of action, 166–7
Certainty, requirement of, 39
Certificate, 49
  effect of, 49
Certification of work, 21
Changes, see Variations
Charges, 47
Circumstances beyond control, 72, 75–7
Commissioning, 23
Company law, 186
Completion of work, 32
Condition, 96
Confidential information, 132
Consequential loss or damage, 117
Consideration, 3, 4, 28
Consumer, dealing as, 119
Consumer credit, 4
Consumer Protection Act (1987), 113
Contract, 1, 2
  collateral, 106, 140, 145
  conditions of, 8–12
  entire, 43
  formation of, 3–18
  legal requirements for, 3, 4
  oral, 11, 14
  simple, 4
  terms of, 8–12, 94
  under seal, 3, 167
Contractor's obligations, 22–4, 85, 87
Contributory negligence, 122
Co-ordination of work, 23
Cost and freight (C&F), 60
Cost, insurance, freight (CIF), 60
Counter-offer, 6, 7, 8, 10, 153
Course of dealing, 14, 15
Credit, 45, 53

192 *Index*

Damages, 98–100
Death and personal injury, 119
Defects, 52, 70, 71, 96
Defences to product liability, 115
Delay, 52, 70–71
Delivery, 57–62
  note, 11
Description of goods, 95
Design responsibility, 36, 89–95
Determination of contract, *see* Termination
Discrepancies, 17, 18, 25
Documents, 175
  of contract, 13, 17, 18, 177
  priority of, 13, 177
Domestic sub-contractors, 147
Drawings, 25, 26, 36, 87

Economic loss, 110–11
Engineer, 19–22, 36
Entire agreement clauses, 13, 14
Entire contracts, *see* Contract
Estimates, 6, 8, 40
European Community Law, 1, 120, 121
EC Directive on Product Liability, 112
Ex quay, 59, 62
Ex ship, 59, 61
Ex works, 59
Examination of site, 42
Exclusion of liability, 116–21, 123–4
Express terms of contract, 94
Extension of time, 76–7, 84–5

Facsimile, 7, 15, 16
Failure of contractor to perform, 154
Fairness, 11, 12, 118–20
Fitness for purpose, 89–98
Fixtures, 130
Force majeure, 72, 75–7, 81–5
Free issue of goods, 141–2
Free on board (FOB), 60
Frustration, 50, 78

Guarantee, 4
  contracts of, 4
  express guarantee of goods, 186
  performance guarantees, *see* Bond

Health and safety, 177
Hire, 11

Identification of goods, 130, 136
Implied terms, 94
Importer for product liability, 113
INCOTERMS, 58–61
Indemnity, 107, 108, 155
Information, giving of, 5, 25
Insolvency, 46, 130, 149, 150
Institution of Chemical Engineers, 183
Institution of Electrical Engineers, 183
Institution of Mechanical Engineers, 183
Insurance, 56, 160
Intellectual property, 131–7
Intent, *see* Letters of intent
Intention to create legal relations, 3
Interest, 47, 48
Invitation to tender, 5
Invitation to treat, 6
Invoice, status of, 67, 136

Joint Contracts Tribunal (JCT), 171, 184

Late delivery, *see* Delay
Latent damage, 169–71
Letter of comfort, 157
Letter of credit, 43, 47, 129
Letter of intent, 5, 8, 9, 12
Liability, 106–24
  in contract, 106–24
  in tort, 106–24
  see also Product liability
Lien, 48
Limitation of actions, 21
Limits upon liability, *see* Exclusion of liability
Liquidation damages, 52, 64–69
Liquidator, 187
Lump sum contracts, 40

Manufacturer, liability of, 108–10, 112–14
Merchantable quality, 95
Misrepresentation, 6
Mistake, 41–2

Negligence, 108–11
Negotiations, 1, 2, 3, 5, 8
Nominated sub-contractor, 144–5, 147

Offer, 7, 8
Ownership, 125–37, 177

Index 193

Payment, 42–54
  non-payment, 48, 49
  stage payment, 43, 44
Penalty, 66–8
Performance, 34, 97
Price, 39–47
  action for price, 9, 53, 167
  firm price, 39
  fixed price, 39
  variable price, 47
Producer, 113
Product liability, 112–15, 123, 168
Programme, 25, 74–5, 87
Property in goods, *see* Ownership
Purchaser's obligations, 24

Quality, terms about, 89–105
Quantum meruit, 188
Quasi contract, 9
Quotation, 6

Reasonable price, sum, 9, 40
  terms of contract, 11, 12
Receiver, 46, 188
Records, 79, 80
Recovery of goods, 130, 135–6, 149
Rejection of goods, 70
Remedies for breach of contract, 85
Representations, 6
Repudiation, 85, 188
Restitution, law of, 9
  principle of, 9
Retention money, 44–6
Retention of title, 128–31, 135–6
Risk, 55–7

Safety, 114
Sample, 95
Scotland, 111
Scots law, 111
Site conditions, 42, 43
Spares, 23
Special conditions of contract, 175
Specification, 24, 175, 176
Standard conditions of contract, 15

Sub-contractor, 138–51
Subrogation, 160–2
Sub-sales, proceeds of, 130
  title to goods, 130
Supplier, product liability of, 113–14
Suspension of work, 49, 81

Telex, acceptance by, 7, 16, 17
Tenders, 5
Termination of contract, 50, 51, 80, 82, 85, 86
Tests, 31–4
Time for completion, 32
  when of the essence, 62
Time for payment, 42–47
  when of the essence, 49, 62
Time limits for action, *see* Limitation of actions
Tort, 107–11
Trust, 46

Unfair Contract Terms Act (1977), 54, 118
United States of America, 112
  laws of, 112
  product liability in, 112
Unsecured credit or debt, 46

Variations, 26–30, 37
Void terms of contract, 66, 119

Waiver, 30, 189
Warnings, 114
Warranties, 96, 100–5, 189
Work and materials, 126, 127
Writing, 4
  requirement for formation of contract, 4
  requirement for variations, 29
Written standard terms of contract, 171
  identification of, 10, 11, 13, 15, 17, 18
  incorporation into contract by common understanding, 15
  incorporation into contract by reference, 15
  oral agreement on basis of, 11

DISCARD